Getting Results: Teaching Math So Everyone Wins

Lessons I Learned Coaching Football to Plan, Teach, and Increase Student Confidence without Assigning Homework

David R. Evans
Middle School Teacher
Assistant Football Coach

Mary K. Naunas, Ed.S
School Psychologist

Published by Dave R. Evans and
Mary K. Naunas.

ISBN: 978-1-7351882-1-8

To Teachers Everywhere

For your dedication, hard work, and doing what's right for our children, despite the obstacles.

Table of Contents:

Introduction i

Chapter 1 1

How Being a Football Coach Made Me A Better Math Teacher and How It Can Work for You

Chapter 2 9

Why Coaching and Math

Chapter 3 17

What Successful Coaches Know and How They Use it to Get Results

Pre-Season: 33

Planning the School Year to Get Results

Chapter 4 35

Starting Your Plan to Get Results: Setting a Goal That No One Thinks You Can Reach

Chapter 5 53

Yearlong Planning: How Many Days Do You Have to Cover the Material Your Students Need to Know...and How Can You Get More?

Chapter 6 69

Prioritizing Lessons: What Do Your Students Need to Know to Get Results?

Practice 85

Teaching Your Students to Be Math Stars

Chapter 7 87

Setting Up Your Playbook to Get Results

Chapter 8 101

Presenting Instruction: Making Every Minute Count

Chapter 9 119

Coaching Your Class: Developing a Winning Team to Get Results for All

Chapter 10 137

The Case for "No Homework"

Game Day 161

Every Lesson is Your Game Day

Chapter 11 163

Coaching Yourself: Getting Your Head in the Game

Chapter 12 173

A Winning Season: and the Best Year of Teaching Ever

Acknowledgements 181

References 185

Introduction

There are about 7.2 million teachers in the United States, according to the 2011-2012 U.S. Census, and while most are confident they know their subject, nearly 75 percent of new teachers and 25 percent of experienced teachers believe they could learn to organize instruction, manage their time, motivate their students, and manage their classrooms more effectively. Are you among them? What if you could improve all those things while watching your students make more progress than you ever thought possible? That's what my brother figured out how to do.

My brother, David R. Evans, a middle school teacher and secondary football coach, came to me a few years ago and told me he was tired of hearing there was nothing to be done about poor math achievement and unmotivated students, on top of completing the full curriculum each year. He had started to notice that the answer to getting results in his classroom was right there on the football field. I worked as a school psychologist for almost thirty years. I consulted with teachers, developed in-service training, and observed in many a classroom. I knew the value of a simple idea that helps guide change. I knew my brother had found just such an idea. In this book, he will show you how to use the time-tested instructional practices and leadership principles the best football coaches use that Dave applied in his math classroom to teach more math in less time while building student confidence along the way. Dave got results, and his students won.

This book provides you with a complete plan for evaluating instructional practices, developing strong working relationships with your students, and coaching yourself in a way that can help you stay focused on student achievement and development. The book, written in a how-to format with examples from Dave's classroom experiences, is designed so you can implement each part of the process while being inspired by one teacher's experience. By using the principles of planning, preparation, goal setting, and mentoring used by football coaches and tested in Mr. Evans's classes, you will learn how to increase student achievement in a high-energy classroom where nearly all students learn strategies and habits that will guide them to success on their next "test day" for years to come.

You'll notice I said "simple" and not "easy." Dave's plan is based on a simple idea that is focused on student success leading to better outcomes. But it's not easy to make that happen. As you use this plan in your classroom, you'll be teaching with greater focus and leading your class with more intensity every day. The effort will be worth it, though. As you read Dave's story, you will learn to:

- Rethink your approach to teaching math so it becomes about the success of your students, not just completing the curriculum (but you'll do that too!).

- Develop student-centered goals no one thinks you can reach.

- Plan effectively, so no time is wasted, using strategies that successful football coaches use to get results under pressure.

- Advocate for the instructional time you need to get results.

- Use teaching methods that help students learn ordinary skills extraordinarily well and then apply them to the right problems at the right time.

- Increase academic engagement so all teaching time is quality time *and homework is no longer necessary.*

- Learn why giving homework can make your job harder.

- Use classroom leadership skills with the focus on building a team with a clear mission based on a common goal.

- Develop a classroom environment that students want to join while decreasing student fatigue and building student confidence and poise under pressure.

We profiled coaches that we feel exemplify the best practices in football coaching and leadership. Educators influenced some in one way or another, but these coaches have the added experience of working under pressure. Dave then shows how their practices and philosophies can be used to develop the tightest plans, most effective instructional strategies, and the highest possible academic engaged time while developing the close working relationship that mathematics students need to become confident learners. But he doesn't stop there. He shows not just how these strategies work but why they work. If asked to justify the use of the strategies in this book, you will be able to do so by referring to the *Science Behind the Success* sections found in the chapters.

Dave and I believe the experiences and principles laid out in this book can benefit most teachers. Read his story and follow the steps in each chapter, and by the end of this book, you will have a plan that focuses on student success while completing all the curriculum you're required to cover in a year. You will also develop a mindset that allows you to channel the panic you often feel into urgency, giving you the energy you need to show your students how to get results. If you're a new teacher, you'll probably want to follow each step to get the most benefit. If you have more experience, you may find ideas that help you fine-tune your practice. Who better than a fellow teacher to understand the stresses you face and show you a solution that gets results?

Chapter 1
How Being a Football Coach Made Me a Better Math Teacher and How It Can Work for You

"Don't mistake activity for achievement—practice it the right way."
John Wooden, Head Basketball Coach, UCLA

"The willingness to experiment with change may be the most essential ingredient to success at anything."
Pat Summitt, Head Women's Basketball Coach, University of Tennessee

That's impossible. There is no way you can get through all the math chapters and get these kids to be A or B students.

And I hadn't even mentioned I wouldn't be giving homework.

That was the reaction of one of my team members when I told my middle school team my plan. I knew when the order came from the top to not only teach all the math curriculum but also teach it for success, I could make it happen. I already had a plan in mind, but now someone had given me a challenge. My pride wouldn't let me walk away from that one.

They said I couldn't possibly take an ordinary classroom full of regular kids and turn them into math champions. They said that I couldn't teach all the required content in the limited time available. They said that I couldn't boost my students' confidence and give them the skills to earn As and Bs.

And as for my additional claim that I could do all that without assigning homework—well, no way. I must be dreaming.

It wasn't just the directive that got me started or the naysayers. I had been teaching for a while and had had enough of weary, discouraged kids. I was tired of hearing there was nothing that could be done about that either. Year after year, middle school students came to my class exhausted and unmotivated. They were convinced they weren't good at math. They had hours of homework ahead of them that they didn't understand. They had parents and tutors at home waiting to go over what I had taught them that day. When they finished their math homework, they still had to do the homework they had for their other subjects. They knew there was no end in sight. Were they pacing themselves with me during classroom instruction, conserving energy for the activities and homework that lay ahead for them?

Yes, I was dreaming, but my dreams eventually came true, and so did the dreams of my students and their parents. My plan worked. I won, and they won, and I'm here to tell the story. With a little help from the world's most successful football coaches, I succeeded. I made the impossible possible. And you can too.

I Realized I Already Had the Skills to Turn My Class Around—I Could Coach My Class

I was a varsity or junior high football coach in my school district for more than twenty years. I loved the game, and I loved coaching kids. I wanted to be the best coach I could be, so I read about the great coaches of pro and college teams, the strategies they used to teach the fundamentals of the game, and how they motivated players. Many went further than the fundamentals and taught their players a deeper understanding of the game as well as the techniques. That was my goal for my math students. I started to realize many of the coaches I read about had come into demoralized organizations and had to deal with players who didn't want to get on board with their plan, and despite all the challenges, they still managed to build great teams. Those are the challenges I faced in September. I noticed that what they were doing was good teaching. And they were doing it under tremendous pressure to get results in short timeframes—every week during football season.

At the same time I was reading coaching books, my sister, a school psychologist, was reading about coaches for a parenting book she was planning. Between us, we found five coaches that we felt highlighted the essentials of good coaching. Their approaches to planning, teaching fundamentals, and helping their players get a deeper understanding of the game and their roles in it was an easy fit for teaching math. We concluded that the best coaches make sure their players can make independent decisions on the field. They develop their players personally so they have the confidence to make decisions both on the field and after they leave the game. The parallels were clear; these were the goals of good teaching too.

The Game Has Changed for Teachers, and Football Coaches Know How You Feel

Laws change and initiatives come and go, but every new initiative brings new challenges for classroom teachers. The urgency teachers have felt since 2001, when the No Child Left Behind Act became law and set high standards requiring all public schools to make adequate yearly progress for all students, is a similar kind of urgency that football coaches at any level feel very acutely. They need to build winning teams with the players they have in the time they have, not the amount of time they want.

Have there been times when you spent a sleepless night wondering how to get it all done? Getting results for our students has never been more important or felt so unattainable.

Teacher standards can feel unreachable, and the public scrutiny while striving to reach these standards adds up to a high-pressure work environment. There never seems to be enough time to get through all the lessons in the curriculum, prepare for the high-stakes assessments, and make sure that all our students will get it when we're done. When you think about the amount of time actually available to teach in a school year, that sense of urgency increases even more. Football coaches know a thing or two about high pressure.

Yes, football coaches know how you feel. They put their jobs on the line every week during football season. Sports writers all over the country second-guess them. Owners want wins each week to fill stadiums. Morale among the players can go south very quickly and sabotage practice if teams don't win. Not to mention that coaches who don't get results will be brushing up their resumes at the end of the season. For football coaches to be successful, they need great teaching strategies. I found great coaching strategies helped me teach math better.

They Are Judged by Results, so They Focus on Results

Reading the papers on Sunday mornings in the fall, you can see football coaches have different ways of dealing with the pressure. Some blame the players. They come to the next practice with a pep talk about trying harder. They institute more drills and watch their players get flat as fatigue sets in.

But the legendary ones already have a plan in mind. They know that they must focus on results, because that's how they are judged. They start the season with a vision for their success and a plan to reach it. They stick to their plan, adjusting their strategy, and communicate this to their players before each game. Long before the first game, though, they teach the fundamentals of the game to mastery and the team's strategy to every member of their team. And they do it on a very tight timeframe to a variety of learners. We don't need to reinvent the wheel to teach under pressure. How do they do this, and how do they do it so successfully? By taking a page out of these coaches' playbooks, we can set up all aspects of the math classroom for optimum success. The rest of this book will show how to get results with tired, unmotivated kids—without assigning homework.

Getting Results

A good coach seems to enjoy the fact that they are accountable.

- They design their program.
- They take responsibility for its success.
- They keep a watchful eye that the program is implemented consistently and correctly.
- They make it clear to everyone on the team what the desired outcome should be.

If things aren't going well, a coach steps in and assumes responsibility. That is their challenge. They want to see their team grow in skills and poise on the field, even though the stakes are high for them professionally. With this kind of role model, I knew I could develop a blueprint for success in my math classroom.

I designed a plan, implemented it in my classroom, and showed myself it could be done. My planning was intense, but it paid off. I reached my goals. By the end of the year, I was very pleased with the results. Almost all my students got an A or a B on every test they took. I put in a long, hard day of intense instruction each and every day, but in the end, my kids learned what they needed without tutors or long hours of homework. They left my class with something in the tank for their next teacher. I kept my expectations high, and together we made success happen for almost every member of my "team." And along the way, my "players" learned to have confidence in their math abilities and in themselves. I never looked at planning, teaching, or classroom management the same way again. I didn't miss grading homework either!

We all bring our personal skills and strengths to teaching; what I learned coaching football can help you plan with more purpose, teach with more focus, and give you the pride of helping your students get their best results gaining much-needed self-confidence along the way. Let me show you how I did it.

How to Use This Book

Have you ever sat in an in-service and thought, "I could be setting up my room right now"? Yeah, so have I. Our schedules are busy. Between the demands of classroom prep, teaching, and the responsibilities of family and after-school activities, every minute we have in a day is precious.

I know you don't have a lot of time to stop and reflect on your planning or class management, and you certainly don't have time to read a textbook or two about it. That's why I've written this book in short chapters that you can start implementing quickly, one step at a time. I know you know how to teach. My purpose is to show you ways to fine-tune the things that can cause frustration by using principles that successful football coaches have been developing for over thirty years to get great results.

A knowledge of sports isn't necessary to fully utilize this book. Enjoy getting to know the coaches whose stories I tell. The principles and practices in relation to the examples set by the coaches will make sense, particularly in building your math students into a winning team. I got the best results when I used all the practices of planning, teaching, and leadership the best coaches use in my classroom, so I suggest you follow the steps in each chapter to build a solid foundation for your class.

By the end of this book, you will have a solid plan that makes the most of all your effort, that plans so kids will get what they need to understand a concept, that finishes all the required content by the end of the school year, and that has the kids in your class working with you instead of disrupting your class. Just follow the chapters step by step. Let's get started.

Chapter 2

Why Coaching and Math?

"Practices are a little like a classroom—you're teaching."
John Gagliardi, Head Football Coach, St. John's University

"Most people get excited about games, but I've got to be excited about practice, because that's my classroom."
Pat Summitt, Head Women's Basketball Coach, University of Tennessee

"Excellence is doing ordinary things extraordinarily well."
John W. Gardner, American Educator

It Wasn't a Compliment...

I was sitting in an in-service one year, and the speaker told the group that football was the best-taught subject in school! It wasn't meant as a compliment. The speaker wasn't a booster for football. The comment was intended as a slam against teachers. But it certainly got my attention. I sat up a little straighter, and I started to think about what I had learned about coaching football from my own experiences and from studying successful football coaches.

Football practice brings together quality aspects of a perfect classroom, I realized. The coaches understand their subject matter (football). They are enthusiastic about teaching it. Coaches have their playbook ready and detailed plans to get everyone ready for game day. The players are highly motivated to learn and are excited about being there. They know they can count on their coach to help them become their best. They see the coach is committed to their success on game day. It sounded like the perfect classroom to me.

But my math class was less than perfect. Unlike football players, who attend practice because they want to, my students were in my class because they had to be, and many of them weren't excited about it. Among any group of students who showed up in my class the first day of school, some hated math, others weren't interested, and some had decided long ago they weren't good at math and were never going to be. Before too many days of school had gone by, my students would be coming to class tired from doing hours of homework the night before with their parents. Why even try to learn when they could tune me out for a class period and then move on with their day?

My math classroom didn't look much like football practice either. It had more interruptions than any football coach has ever had to contend with. The phone would ring; someone in the office had a question for me. Announcements would come in over the loudspeakers. Students would act up in class. Lots of differences and problems interrupted our time in the classroom. Still, I wondered. Could I use what I learned coaching football in my math classes? Could my students get results on test day, just like football players get results on game day, using the same principles all successful coaches seemed to know?

I thought about the successful coaches that I had read about and whose ideas and inspiration had helped me be a better football coach. They weren't perfect people. They had flaws and made mistakes, but their dedication to their craft and determination to be the best coaches they could be inspired me. I was interested in their experiences and seeing what I could learn from them. How did Bill Walsh make the 49ers the team to beat in the 80s? How did Dennis Creehan, the expert on the Wing-T offense, turn around pro and college teams, often in one season? How did John

Gagliardi put together winning seasons for the Johnnies year after year for over sixty years? What was Glen Mason's focus when faced with the job of turning around teams at the University of Kansas and the University of Minnesota? How did Tony Dungy get the Tampa Bay Buccaneers back on a winning track? As I got to know their stories, I became a better football coach. Now I was going to see if those skills could improve my teaching and classroom leadership.

Making the Jump from Football to Math

After considering all the pressures that successful coaches face and how they address them on the practice field, turning their teams around, I could see how the principles I had been learning and using after school to coach football were actually excellent teaching strategies but wasn't sure that was enough. I started thinking about the similarities between teaching football and teaching math. There are similarities in content. There's a focus on fundamentals. Those fundamentals must be taught to mastery. The best coaches also teach a deeper understanding of the game so their players can make decisions on the field. Isn't that what I wanted for my math students too?

The more it thought about it, I realized that there was a close match between what I needed to teach my football players and the content I had to teach to my math students.

- *Fundamental Skills*

In both football and math, fundamental skills need to be learned to mastery. Just as football players need to understand skills like blocking, math students can't progress without understanding addition, for example.

- *Errors*

In football, players can't make errors, or the play falls apart. In math, students can't make errors, or they don't get the right answer.

- *Strategies*

Once the fundamentals are mastered, the challenge is not over. Players and students also need to have strategies to solve problems and know when to apply them. Football players, for example, need to know how to find an opening to run the ball just like math students need to learn to break down word problems. They both need the big picture, or a deeper understanding of the game, to be independent learners.

My classroom wasn't functioning as well as my team. Once I realized just how much coaching football and teaching math had in common, I realized I had the tools to turn my math classes around.

Strong Focus on Fundamentals: Doing Ordinary Things Extraordinarily Well

In both football and math, fundamental skills must be mastered before a real understanding of the game can happen, so this needs to be a priority when setting up instruction. In both football and math, the more active engagement, or hands-on time, is provided, the more students and players learn. They need to be fully engaged for these fundamentals to be taught in the least amount of time possible. Why? Because game day and test day are coming fast.

On the football field, practice is a busy place. Coaches require players to repeat skills until they can do the fundamentals to mastery. Corrective feedback is given quickly so more effective practice can continue. The more time players spend practicing skills, the more likely they are to master them and get results on game day. Coaches know that time must be planned into the practice schedule to ensure that mastery occurs. Once the players do ordinary things extraordinarily well, they can concentrate on the big picture, learning the strategies needed to win. If I was to model my teaching methods after these coaches, then I had to have time for individual practice as well as group instruction. Math students learn by doing too, so I needed to give group instruction in short bursts and provide opportunities for my students to practice the fundamental skills. Following these strategies, I would also have to provide corrective feedback to students frequently while they were learning those skills so they got the most benefit from their reps, or repetitions, and not get stuck or practice a skill incorrectly. To get the most out of class time, I would need to give that feedback quickly and not take too much time away from practice.

Math and Football Instruction Are Sequential in Nature

Football practices follow a sequential process from working on individual skills to coordinating their skills with other team members to do a successful play. This makes it easy to see where the process breaks down and instruction needs to begin. Standing on the field with the other football coaches at practices, I watched our players work through drills. It was easy to see what skills I needed to teach just by watching the players practice. Analysis of the strengths and weaknesses of the players works the same way. Each player had weak areas that needed to be remediated before the next game. In math, fundamental skills and processes are also sequential. By observing them in class, I could see what skills my students needed and set up instruction and practice so they could master those fundamental skills. I would need to find time to make sure everyone in my class had the skills to succeed on test day.

Need a Big Picture for Understanding

To become the best they could be, I noticed, both football players and math students must not only master fundamentals, but they must also know when to use them. I learned that both football coaches and math teachers must keep both these goals in mind when working with their students. Understanding without mastery of fundamental skills leads to inconsistent performance. On the field, if a key player doesn't do their job, the points are lost. In the classroom, students who don't have math fundamentals mastered struggle to solve real problems, like the ones they will see on the next test, even if they know the process or can follow the logic.

On the football field, the play can end badly if players don't know what to do if things don't go as planned. By instructing them to see the objective of the play and helping them develop strategies for dealing with possible situations they may face, they are more likely to get results on game day. The coaches I read about asked questions during instruction to help bring out their players' problem-solving. Questions like "What would you do if?" get players thinking about ways to be successful on game day. This increases their independence on the field. I would need to do the same thing in my math classes, building student confidence by teaching them strategies to address the situations they could face on test day.

To have time to teach both fundamentals and strategy, teachers and football coaches can't waste a minute or lose focus on what needs to be instructed. Drills must be efficient. Practicing fundamentals in the context of situations helps students and players understand the big picture and is an efficient use of time during practice or in the classroom. Running drills in the context of the play helps all team members understand why they are learning the fundamental skills being taught that day and their role in helping the team win. In math, students drill fundamentals with the understanding of why that skill is important to solving problems on the next test. If I helped them learn to judge which skills to use in what situations, my students would get a deeper understanding of math problem-solving and be able to choose the best strategies for solving problems on the next test.

Once I realized the similarities math and football had in content, I knew I could plan, teach, and coach my kids to get results on test day, just like I got my football players ready for game day. My reading of successful football coaches was about to be put to the test in my math classes. The most successful coaches in pro and college football all seemed to know how to turn around failing franchises, staying confident and focused on their goals. They were determined to get results regardless of the critics who said it couldn't be done.

I would use the lessons I learned from coaches to deal with pressure, plan successful lessons to fit the time I had, run an active "practice" during class time, and, most importantly, revise my leadership to match coaching. I was sure that I would be able to develop a plan, teach the plan, and keep expectations high for my math students. The rest of this book is set up to help you develop the mindset that successful coaches use to get results and how to prepare, plan, provide quality instruction, and lead your students to get results.

Chapter 3

What Successful Coaches Know and How They Use It to Get Results

"The superior man blames himself. The inferior man blames others."
Don Shula, NFL Coach

"A man can make mistakes, but he isn't a failure until he starts blaming someone else."
Sam Rutigliano, NFL and College Football Coach

The last play ends; the crowd cheers. Friday's game is over. The team won, and they are pumped. Fundamentals were executed flawlessly. There was no hesitation when the opposing team threw their best play. Their preparation paid off—this time. But the next game day is coming up fast.

For the coach, preparation for the next game has already begun. It starts as they and the players walk off the field. Injury reports and observations during the game about what worked and what didn't run through the coach's mind. Next week's team will have different strengths and weaknesses. Will strategies need to change? Will practice need to be adjusted to put the players in a position to win?

The coach analyzes their impressions from the last game. Now they need data to back up their conclusions. That starts with film study. Saturday morning, the team will come together for film study. Film study is much like going over a test. It's a time for players to see with their own eyes if they have been performing their tasks at a high level. Every player can see if they performed their task right or wrong; the film doesn't lie. It's a motivation to improve for next week. No one wants to see themselves on film performing badly.

Sunday is a day off for players, but not for the coach. They spend the day getting to know next week's opponent. Coaches in charge of the offense are watching film of their next opponent's defense to identify ways to attack their best plays. Coaches for the defense watch the film of the opponent's offensive plays. They use this information to set up their plan to stop the other team's best plays. They will bring their ideas to Monday's meeting.

On Monday, the coaches consider all the data they collected from film study and their observations of their players' skills. With an eye for what their players know, they develop their strategy for the next game. They need to prepare for the what-ifs. They must look at problems and find ways to solve them. All solutions to problems must be solved by answering the following questions:

- Will the solution work against the opponent?
- Can it be taught?
- Can the players execute the plan on game day?

Some plans might work in theory, but it has to fit your team's talents. Maybe the players could do it, but you don't have the time to teach it. Solutions need to be practical and efficient, because the timeframe before the game is short. Football coaches know the reality of their situation and plan practical solutions.

After a light practice on Monday, it's time to install the game plan Tuesday. Some of the most intense teaching occurs on a Tuesday. The coach knows that the next week's game will be won on the practice field. Most teams practice many game situations on this day as the coach watches closely to find out what works and what needs to be eliminated from the game plan.

Coaches use Wednesday's practice to make adjustments to the practice plan based on what they learned during Tuesday's practice. Key strategies are reviewed. Fundamentals are practiced until they are done consistently. Coaches try to develop confidence in the players. To do this, they must be careful to take their players through each step of each play without introducing too much new information at one time. There is less live contact today, but coaches continue to install the game plan for the offense, defense, and special teams, making sure their players run the plays correctly each time and understand what to do in each situation. To make sure it will go well, the entire practice will be planned out for every minute. There will be no time wasted on thinking about what play to run or what drill to do. Players will hustle from one group to the next throughout the practice, so they get their reps in. Football has been called a game of mistakes. The team that makes the fewest mistakes usually wins.

Thursday, for most high school programs, is pregame preparation. Players don't run too much so they can get their legs back for the game. The focus is on getting them ready mentally. The kids need to be sharp and think about their job and not about the hype of the game. Players should leave the field feeling physically strong and mentally confident.

Friday dawns on game day. Most coaches have a plan for every minute leading up to game time. After a pregame meal, for example, the players will come to the locker room or field house to get ready for the game. Most teams have a walk-through as part of their pregame, which allows coaches to remind the players of the most important parts of the game plan one last time. All players need to suit up, warm up, and get mentally ready to perform at their best. Pregame activities should make players feel they are at their best and ready to play, not exhausted and discouraged. Now it's up to them. The coaches take their place on the sidelines and watch their players get results.

The Principles of Successful Coaching

There are six basic principles that I learned from studying successful coaches and from my own experiences as a football coach that guided me in my classroom to help my students become more successful than anyone thought was possible. These principles make up the vision for success I used while coaching football and teaching my math class and, as we will see in the chapters ahead, will help you prepare for a successful school year.

Successful coaches:

1. are accountable for the results of their team;
2. have a vision for getting results and set goals based on the vision;
3. develop an effective plan to reach their vision;
4. communicate the vision, and their plan to reach it, to their players;
5. implement their plan using effective teaching methods for fundamentals and application; and
6. develop the full potential of players through personal input and coaching.

Successful Coaches Are Accountable for the Results of Their Team

Every good coach understands that they are fully responsible for the team's results on game day. A special teams coach dies a little inside if the other team runs a kickback for a touchdown. If a linebacker misses a tackle, the linebacker coach feels the pain.

It's a sign of weakness to blame the staff or players when they lose, though. Head coaches must take full responsibility for the performance of their players. After all, their jobs are on the line every week, which is obvious when headlines appear in the sports section of the newspaper during football season. A head coach is under tremendous pressure to get results. But the satisfaction of being successful drives them to strive for excellence and to be a role model for their players. When the team wins, it's the coach's victory too.

Always aware that they are accountable for their program, how do successful football coaches judge themselves and gauge the progress of their team? Although wins and losses are important, building a winning team requires more. They must evaluate the factors that make winning possible. So, successful coaches ask the following questions:

- Did I put my players in a position to win?
- Were they prepared for what could happen in the game?
- Did they perform the tasks the way they were taught?

Coaches know they can't just look like they are doing it right, they must get it right. When things aren't going well during the game, successful coaches look at themselves and say, "This is my fault. What am I going to do the change this?" They then analyze what went wrong and begin to plan how they will address the problem. As soon as the next practice, good coaches follow up with the team, breaking down film and giving their players a concrete plan for practice that week to fix the problem. That plan may involve reteaching fundamental skills or helping their players read situations better so they use the right skills and knowledge at the right time to be successful on the next game day.

Successful coaches understand they are accountable for the success of their team. They design their program based on their vision and the team's needs assessment. It includes a system for getting results and sets high expectations for everyone on the team. They make it clear to everyone on the team what the desired outcome should be. If things are not going well, they step in, assume responsibility, and address the problem. That is their challenge. They want to see their team grow in skills and poise on the field.

Successful Coaches Have a Vision for Getting Results and Set Goals Based on the Vision

Getting results starts with the coach's vision for success and concrete goals to make the team successful on game day. When I think of Bill Walsh, I think "planning for situations." His books, *Building a Champion: On Football and the Making of the 49ers* and *Finding the Winning Edge,* inspired me and provided me with a plan to be successful as a coach and as a teacher. Named NFL Football Coach of the Decade for the 1980s by the Pro Football Hall of Fame, Bill Walsh is best known as head coach for the San Francisco 49ers. He took the demoralized, losing, least respected team in the NFL and turned it into a winning franchise that set an NFL record of fourteen straight seasons of at least ten wins and a postseason record winning ten of his fourteen games and six divisional titles, three NFC championship titles, and three Super Bowls. His election to the Pro Football Hall of Fame is well deserved.

In 1979, when Walsh took over a demoralized 49ers franchise, only he could see a better team. His vision for turning the team around was clear in his mind, and he broke it down into the steps needed to make it happen. Then it was up to him to communicate his vision clearly to his players, many of whom didn't want to listen. He also had to have the determination to make sure everyone followed his plan. His leadership is the reason for the complete turnaround of the team. He was determined to put his team in a position to win, and his commitment to his vision got results (Walsh, 1990).

Visions are just dreams until goals are set. In sports, the goal is obvious. The team needs to win as often as possible. For the coaches introduced in this book, that goal set a very high expectation that very few, other than themselves, probably thought was possible. Their visions for their teams and the goals they set to reach it focused their planning and their instruction on what was important: getting results. Then they set ambitious goals no one thought they could reach and stuck to them. Their goals reflected high expectations for all players, and they never lowered those expectations.

Each coach's vision sets the priorities for what players work on in practice to win on game day. Bill Walsh focused on mastery of fundamental skills and developed very detailed plans for practicing them, so every player was ready to execute each play flawlessly on game day. Priorities are set based on what the coach sees as the necessary skills and strategies that must be taught to mastery for the team to win. Like Walsh, all the coaches I'll talk about got results by following their personal vision and getting their players to follow it too.

Successful Coaches Develop an Effective Plan to Reach Their Vision

All successful coaches assess the needs of their team and create a system for success based on that assessment. A top coach has a great system and develops a deep understanding of the system so they are prepared to answer any questions their players have about its effectiveness and how it will help the team meet its goals. Some develop it themselves. Others adopt one they feel will get results. For all football coaches, developing a system for success includes deciding what offensive and defensive systems will be in place.

Once they decide on a system, the coach must understand it, know how to teach it, and decide what skills must be taught during each practice so the team can get results on every game day. For Bill Walsh, this was the West Coast offense, for which he became famous. Using brains over brawn, he put his players in a position to win. Walsh used detailed planning to eliminate panic during a game. Every player learned what was needed to get results against that week's opponent. They practiced every skill needed to make each play work until they could consistently execute every play in Walsh's playbook for game day. By sticking to the plan, decisions on game day could be made with confidence, making it less likely that panic would set in. If something went wrong, he wouldn't abandon the plan and try to make up something on the fly; he would stick to his plan. His players knew that if something didn't work, they would hear about it at the next coaching meeting and practice would be adjusted to improve the team's performance next time (Walsh,1998).

Planning in football is not just for game day. There is also a plan for every practice. In his book *Finding the Winning Edge,* Walsh's system is laid out with very detailed lesson plans, including what would be covered in every practice throughout the season with explanations for why those skills were needed to get results. His objectives were clear, and everyone understood them. Just like on game day, they stuck to the plan during practice too. Walsh and his coaches walked onto the practice field with a plan already developed during the offseason with the skills and training plans needed for every play the players would run on game day. There would be no deviation from the lesson plan. By sticking to his plan, Walsh never wasted a practice (Walsh, 1998).

Successful Coaches Communicate Their Vision, and Their Plan to Reach it, to Their Players

Once coaches have a system that works and have developed their confidence in it, they need to communicate their vision to the team. They must show their players how to carry out their system to get results. Successful coaches know how to run practices, scout opponents, and perform on game day, among other skills, to achieve results. They have a passion for the game and a vision for winning. But this alone will not make a winning team or even a good one. They must be great communicators. Now coaches find out if they can truly lead.

Leadership starts by communicating the vision to the team so the team will follow the vision. During every practice, successful coaches work toward making sure all their players see how the system can help them get results. They plan how to teach it in ways their players will understand, with some even considering the learning styles of their players. Good coaches are genuinely interested in their players' development, and contribution to the team and want them to feel supported. To communicate in a way that shows players they are supported takes enthusiasm and frequent feedback. For example, Glen Mason, a coach at the University of Minnesota between 1997 and 2006, inherited a football club that had been losing for six seasons. He felt it was important to communicate to his players the ways they were making progress toward their goals in each practice. After doing a needs assessment, he began building on the positives that allowed his players to see their improvement and develop the confidence they needed to keep working to get better.

The coach's actions also communicate leadership. Good coaches are positive and set high standards for everyone on the team. They have confidence in their system and apply it to any problem or skill that needs to be taught. Coaches understand that problems will arise. They know they can't lose faith in themselves just because they can't control everything. When things go wrong, successful coaches go back to their vision, their core philosophy, and the system they built to find the solutions to those problems. By watching their coach stick to his plan, players develop confidence that the system will make them successful. Calm leadership focused on finding solutions using the plan, not feelings, gets results. Once a coach has laid the groundwork for a sound system and a plan to be successful, they need to have the confidence to allow success to occur. They can't panic or allow their players to panic. Players look to the coaches in times of adversity. They need to see a steady hand at the helm.

Successful Coaches Implement Their Plan Using Effective Teaching Methods for Fundamentals and Application

For coaches, games are won or lost at practice. Their plan for implementing their system sets high expectations for all and must be taught in a week. Decisions are made long before the first practice about what will be taught based on the skills needed to get results. A needs assessment of the players' performance helps pave the way. The skills practiced, and the way practice is run, reflects the coach's vision for success.

Successful coaches know they have only a limited number of practices to get their team ready. They need to walk onto the field on game day with a clear understanding of what needs to be accomplished and a plan in place to make it happen. If they don't stick to the plan, time that is needed for instruction will be wasted. Coaches have a sense of urgency to complete the goals for each practice. They stick to their practice plan, so urgency won't become panic. They also need to teach the plan effectively with lots of reps, so mastery of fundamental skills happens by the end of the session.

For a team to get results on game day, they need to execute the fundamentals of the game correctly every time. This requires a mastery of fundamental skills, which are best learned hands-on. Consider Dennis Creehan, for example, who has held many college and pro football coaching positions in his career but is best known as an expert on the Wing-T offense, which is a system of offense used by high school, college, and pro teams. Creehan has written extensively about this offense and how to implement it effectively, which includes how to run an effective practice. In his book series *The Wing-T from A to Z,* I learned how to run a "racehorse," or fast-paced practice, to ensure players know what is expected and get enough reps in practice to master fundamentals needed for game day. In a racehorse practice, the coach and his players work with speed and a sense of urgency to get in as many reps as possible (Creehan, 2000).

Successful coaches also want their players to have a deeper understanding of the coaches' system and the players' place in it. For sixty years, John Gagliardi, legendary coach of the St. John's University "Johnnies,", set up his system for players to learn what worked and what didn't by practicing plays. He wanted to teach independence, which was important in his system because his players decided what plays to run on game days. Gagliardi believed in learning by doing (Collison, 2001). Practices were short, about ninety minutes. He didn't think his players could pay attention longer than that, so he had to make every minute count. He didn't waste time on drills that didn't teach players what they needed for game day. His players got the same amount of physical activity, but Gagliardi used the players' time to develop a plan and to understand their role in the next week's game. Jim Collison, author of the book *NO-How Coaching: Strategies for Winning in Sports and Business from the Coach Who Says "No!"*, believes this is the crux of

Gagliardi's unique approach and his belief of the importance of teaching in a high-performance environment. Gagliardi had no busywork, frills, or timewasters in his program. If it wouldn't help win the game, it wasn't part of his practice. The team's energy, and his, was focused on what was needed to win (Collison, 2001). He kept the players focused on the skills and activities that would increase their chances of success on game day.

"Football is a game of mistakes," according to Gagliardi in *NO-How Coaching*, "and somehow you have to eliminate 'em" (Collison, 2001). To win, he believed the fundamentals must be flawless, so they were practiced frequently but always within the context of the plays that would be used on game day. During practice, teams reviewed successful strategies against their next opponent and developed a game plan. Then they went out on the field and practiced the plan over and over to see what worked and what didn't (Weeres, 2010). They learned by doing, gaining the experience that comes from following their decisions through to the outcome. The players decided what was working and what wasn't, and their preparation gave them both the skills and confidence to play their own game.

By emphasizing fundamentals and game-day situations, these coaches were teaching for success on game day.

Successful Coaches Develop the Full Potential of Players through Personal Input and Coaching

Players need to know that their coach is in charge, has a plan for success, and has a place for them in the plan. They need to know their coach is committed to their success. What makes Walsh stand out among coaches is his ability to build teams rather than collect the most stars. Other winning coaches counted on great talent to get results. Bill Walsh taught all his players how to be their best and developed stars in the process (Walsh,1990). Gagliardi treated all his players, regardless of their talent, as though they could be the best. As far as he was concerned, you never knew who was going to shine, but by including them as full team members, you put them in a position to develop into the best player they could be. It is the coach's responsibility to bring out the best in every player (Weeres, 2010).

The best coaches set up a results-oriented program that allows players to succeed. They do not teach in a sink-or-swim situation. They create an environment for success. Players feel part of something they will build together with the coach by preparing to reach a tough goal together. Relationships develop as the coach and each player get ready for game day. The coaches I respect communicate with their players in an open, honest way that provides a connection with each player. They do this through the context of preparing for game day, however. There's no time for chatting if the team is going to be ready to win on game day. When the players and the coach prepare to reach a tough goal together, they become a team.

Coaches set the tone for how their team will treat each other through both their actions and by their rules. Their personalities dictate how they do this. John Gagliardi used the Golden Rule—treat others the way you wish to be treated—for all relationships on his team, including his relationships with his players. Every player was respected and challenged to improve by asking them how they would address a problem on the field. The same rule applied to their relationships with each other. Gagliardi's more experienced players were expected to help the less experienced ones become their best (Weeres, 2010).

Respect starts with the example set by their coach. Getting ready for game day requires intense instruction and practice in a short timeframe. Mistakes are clear to see and need to be corrected. If they are going to help the team succeed, each player needs to improve quickly. Coaches get to know each players' strengths and problem areas by continually observing them at practice. Timely corrections and feedback given on the practice field help mastery of necessary skills occur efficiently. Coach Creehan, we will see, wants that feedback to happen in a few seconds so it doesn't slow down practice (Creehan, 2000). Feedback is given in a businesslike way, sticking to the necessary information needed to improve without taking away a player's pride. The coach's respect for his players needs to come through, especially if he wants it to carry over to the players' relationships with each other.

Teaching Math like a Football Coach

Successful coaches provide leadership in a very personal way: their way. All great teams begin with the head coach's leadership.

- They are in charge, and everybody knows it.
- Expectations are set high. If the team is to be successful, everyone needs to be at their best.
- Successful coaches put their players in a position to win by preparing them for what they will face on game day.
- Coaches must accept accountability for the results of their team. They must develop an effective plan that gets results.
- They must communicate their vision to their staff and players, setting high expectations for all.
- They must implement that plan so their players learn to do ordinary things extraordinarily well.
- Fundamentals are taught to mastery; nothing is left to chance.
- Finally, and perhaps most importantly, they make sure all players reach their full potential

> so they can make the judgments on the field that win football games.

Teaching math is also very personal: it's your way. In the chapters that follow, I lay out my vision, the goals I set based on that vision, how I prepared for that first day of class, and my techniques that led my class to success. Teaching fundamentals and how to apply them using a hands-on approach in a way that reduces fatigue and makes it clear to everyone in your class that they are successful, not because you adjusted the material to make it easier but because they were able to learn the hard stuff, is a tall order.

This system will have many components, and I'll go through them step-by-step in the chapters that follow. I'll teach you how to break down your year so that you have time to teach everything that's expected by setting priorities based on what your students need most to get results on test day. Because even the best plans won't work if everyone isn't on board, I'll also show you how to develop a team approach to classroom management. Using these principles as a guide, you'll learn how to develop the plan, teach the plan, and keep expectations high. Your students will thank you for it, just like mine thanked me.

Pre-Season:

Planning the School Year to Get Results

Chapter 4

Starting Your Plan to Get Results: Setting a Goal that No One Thinks You Can Reach

"[A good teacher] helps everybody earn an A."

Tony Dungy, former Head Coach, of the Tampa Bay Buccaneers and the Indianapolis Colts

"You must have long-range goals to keep you from being frustrated by short-range failure."
Charles C. Noble, American Civil Engineer

Here's what I learned that all good coaches seem to know: always move forward by setting goals. Good football coaches set goals. While some goals may be for the season, some are quite short: just for today, or even for the next five minutes. In practice, a coach may have a goal to teach a skill in five minutes. If they don't have the time, then some other part of the preparation will be cut short. In a game, a goal for a play may be to gain at least four yards. A goal for a second down might be two downs to gain six yards or three yards per play. Next goal: first down, not a touchdown, not the whole game. Coaches learn to be patient and make it happen. It all starts with a vision and the goals you set to reach it.

Coaches set goals for each season, each day of practice, even each minute of each game. As long or as short as those goals are, each goal needs to be part of a larger vision. The football coaches who inspired me each had a vision for success and a plan to reach it. Many of our coaches faced real uphill battles to make that happen. They had to turn around failing franchises and deal with surly players who were not interested in learning their new coach's system, players' defeatist attitudes, and their own discouragement in facing all the problems they had to solve to be successful. How did they keep from panicking or just giving up? Their vision for their team and the goals they set to reach it focused their planning and their instruction on what was important: getting results. Then they set ambitious goals no one thought they could reach—and stuck to them.

In the next three chapters, I'll lay out my plan for getting results. Like my favorite football coaches, I had to stay focused, keeping my goals front and center. In each chapter going forward, I'll show you how you can get results using the leadership skills of the successful coaches who inspired me and the educational experts who influenced me. At the end of each chapter, there will be a section to help you start your own plan to get results.

This chapter will show you how to start developing your vision and setting goals for the year that will focus your planning and how your goals can be specific enough that even your students will know when they get results. Once you set your goals for your class, it will drive every decision you make as you start planning the year. It will inform every choice you make once you start to teach. It will give you a sense of urgency when making those thousands of decisions you make every day, choosing the ones that will help you and your students get results on test day and ignoring what doesn't.

Setting Goals Starts with Your Vision: What do you want to accomplish this year?

Establishing a vision for my students' performance was the turning point for me and it took me less than a minute to do it. It's easy to get a mental block about what is possible when past solutions haven't worked. I was tired of being told that teachers couldn't get through the whole curriculum in a school year or have all of their students be successful. I didn't want to accept things the way they were any more. I was ready to change how I taught math to make both of those things happen—and I wasn't going to be talked out of it by anyone. With the support of my team, I would make sure *all* my students got results on test day. Instead of complaining about the roadblocks that were in my path, I would set up plans to overcome them.

My planning for that first year became laser focused, dictated by my vision: teach the complete math curriculum to every student in my classes so they could get results on test day all on their own. I wanted my students to be successful, independent learners with skills and strategies that would make them successful this year and every year after.

Now it was time to put my vision into action by setting goals. Football coaches set specific, ambitious goals for the season for their players, and for themselves. They know what's possible and set very specific goals to make their vision happen despite the obstacles. Looking at each daily problem and decision that needed to be made in light of my vision kept me focused. I soon realized the direction of my planning had to change, from curriculum content to student mastery.

I set two goals for myself. The first was to complete the whole math curriculum. This goal, required by my principal, was a no-brainer. The problem, of course, was that not all the kids in my heterogeneous classes were on grade level in math or had the same ability to be successful. Some had been struggling with math for years. Not everyone was going to be ready to get results if I just took the curriculum guide and divided it across the year, setting the pace and not considering my students' skill levels.

Setting a goal to finish the curriculum in a school year wasn't going to be enough. When they left me at the end of the year, I wanted my students to have strong fundamentals and a deeper understanding of the math skills and strategies and be able to apply them in the future, both on tests and on practical problems, both in my class and in other subjects, like science. I wasn't going to let some kids in my classes get it and let the others struggle. *All* my students would get results every test day.

Your Goals Should Tell You When You're Getting Results

I set goals for my students based on the principles I learned coaching and the educational research of Tomlinson and McTighe. In their book *Integrating Differentiated Instruction and Understanding by Design*, Carol Ann Tomlinson and Jay McTighe tell us goals are set to help teachers plan effectively to increase student achievement. These goals are long term, student centered, and start with deciding what the desired results should be. I decided that the desired results for my students were to acquire strong fundamental skills and to gain a deeper understanding of the concepts and systems of math needed to get results on test day.

Desired results, however, need to be measurable. How will you know when this goal is reached? By using acceptable evidence, says Tomlinson and McTighe (2006). What is acceptable evidence? It's when students can complete an authentic performance task. That is, they need to apply what they learned to real-life problems or find solutions to test problems they devise independently. In my case, the acceptable evidence was my students' grades on the tests and quizzes they took throughout the year. I decided that all students would get 100 percent on each test by applying the skills and strategies they would practice in class.

Goal setting is the first step in backward design. Think of it as putting your destination on a roadmap. In this approach, teachers identify the desired results, or destination, then decide how they will know when the destination is reached. The goals are student centered, or the skills they need for success. Using backward design helped me stay focused on what my students would need to be successful.

Goals Need to Be Specific, Have a Time Target, and Be Checked Frequently

Setting the goal that all my students would get 100 percent on every test did what a good goal should. It gave me a specific, measurable target. It also gave me a very clear timeline to reach it. I had one school year to teach all the material in the curriculum to my students so they would learn fundamental skills to mastery and develop a deeper understanding of math principles. I would teach my students to do ordinary things extraordinarily well and develop a deeper understanding of my system so they would be successful on test day.

My goals also informed my instructional decisions. Instead of focusing on getting through the material by the end of the year, my goals gave me a sense of urgency to make sure that my lessons were sharp and provided guided practice and challenges under my watchful eye each day. My goals made me rethink what to teach and how to teach it so my students would be ready to reach the goal each test day. Planning and teaching took on a laser focus when my goals were ambitious.

Goals Need to Be Student Centered

If my plan had been just to get through the curriculum, I could have set this goal for myself and just gotten started. But it wouldn't have been student centered. Every other year I taught before, the kids who got behind had their parents and tutors to help them. Goals focused on only finishing curriculum assume some kids will not do as well as others, but I wanted *all* my students to succeed. Football coaches think about the ability level of each player and make sure they are up to speed so the team can get results. Many of the coaches we talk about took a long, hard look at their players to find out their strengths and skills and what they wanted them to learn to be better team members. For example, John Gagliardi observed players in practice to see if they were making correct decisions, then, using his mentoring style, guided them to improve skills while practicing plays. A weak link keeps the team from executing their plays effectively. Practice plans start with the end in mind and build each players' skills so the team's goal is reached. Coaches ask themselves the question, "How can I put my players in a position to win?"

My vision was very clear. I knew what I wanted my students to be able to do by the end of the year. I would make sure they mastered fundamentals so each student could do

> ***The Science Behind the SuccessThe***
>
> **Mastery or Meaning. Which is More Important?**
>
> Learning skills that need to become automatic, like mastery of basic skills, and developing meaning, or a deeper understanding of concepts, according to cognitive psychologists, are equally important to improving cognition (Bruning, et.al 2011).

ordinary things extraordinarily well. But they would also be able to apply fundamentals to solve problems based on their deeper understanding of the principles and strategies of math. I wanted independent learners when I was done. The goal wasn't how I presented the material. It was the learning of each of my students.

In short, I had three goals:

1. All of the students learn the whole curriculum.
2. All of the students receive a perfect score on every test and quiz.
3. All of the students leave the classroom at the end of the year as independent learners.

Setting my goal of making sure all my students got 100 percent on every quiz and common assessment kept me from looking at certain students as people who couldn't succeed and let me look at them the way coaches look at athletes. I was going to have to ask the question, "How can this student help us win on test day?" My focus changed. I wanted to start looking at each of my students and ask myself, "How can I help this student get 100 percent on the next test?" As we'll see in later chapters, my students became a team along the way to reaching this goal.

Why Goals Need to Be Ambitious

Many experts on goal setting will tell you that goals need to realistic. They need to be something you realistically can reach. If you are being evaluated based on your goals, that is good advice. But football coaches look at this part of goal setting differently. They set very ambitious goals because their job is on the line for them anyway, so they have nothing to lose by setting a goal no one thinks they can reach.

Ambitious goals give coaches a sense of urgency that carries over to every decision they make. Keeping their goals front and center helps them stay on track when setting priorities on what skills they should teach in the time they have each practice. They don't waste time on instructional activities that won't help their team get results on game day. In the same way, planning and teaching took on a laser focus when my goals were ambitious. The main goal, for all my students to get 100 percent on every test and quiz, was very ambitious. Most people will tell you that goal is not possible. I would agree with that. But I still kept it as my goal. Setting an ambitious goal helped me pull hard in the right direction by keeping my expectations high for all my students. Having a goal no one thinks you can reach focuses your planning on what is important by making your priorities clear.

Making ambitious goals also ensures that expectations stay high for all. Every player needs to play well to help the team be successful. When they are playing at the top of their game, the team gets results. That is true for students too. Setting high expectations for all my students helped me ensure that everyone would master the fundamentals and be successful on test day. Just like football, math is a game of mistakes. The fewer you make, the more likely you are to get results. The better your players understand the plays and their part in them, the better they are going to be at executing basic moves and making decisions on the field. Strong fundamentals backed up with a deeper understanding of the coach's system gets results. I wanted the same to happen in my classroom. Strong fundamentals would reduce mistakes on test day.

Sharing Your Vision: How Will Your Students Know When They Get Results?

Your Goal Should Be Clear to Your Students

I soon found out that the goal of getting 100 percent on every test had a second benefit that is rarely considered when developing goals. This goal proved to my students that they were capable of learning something they had found very difficult in the past: math concepts and strategies. To take on classroom after classroom of defeated, confused students who were spending far too many hours per day attempting to get results without success can make it hard to convince the students that they are making progress. But now both my students and I knew when we were successful and could be on a regular basis. My focus was on increasing student achievement in a way every student could see and building confident learners through my classroom leadership. I wanted to show my students they were good at math every day in class, using authentic assessment built into instruction, as well as on test day.

Students need to see their success to believe it. It helps them realize they are making progress as a result of their hard work in class. Seeing their grades improve helps them focus on their progress. Without these objective markers, students will find their own less objective markers and will likely find themselves lacking. Think back to the last time you took a tough class. How did you judge your progress? If you're like most people, you checked to see how the other students in the class were doing and found that one student you knew you would never catch up to. Students very often are in the same boat when they come to math class, and they check their progress the same way: they compare themselves to the smartest kid in the class. If they don't have objective data to show them their progress, they will feel like they are always behind. They feel they will never be able to do what those other select students are doing, so why try? By setting up ambitious goals with evidence they can see, you give them a way to know they are successful. Everybody will get the same test. Everybody will see their results and know they are successful.

Setting an ambitious goal and sharing their results with them got my students' attention. It was also the first step to getting their cooperation. I was going to need it too. The instruction was going to be quick, intense, and focused, and my students were going to have to stay with me every step of the way. I will talk more about getting your students on board through your classroom leadership in later chapters, but the important point to remember right now is that your students must have a way to see they are making progress. You can't just tell them; you have to prove it to them. Discouraged kids have heard it all before. Their teachers in previous years, their tutors, and their parents have told them many times they can do it if they just try hard enough. You need to show them they *are* successful.

Your Goals Will Focus Your Planning

Setting ambitious goals will have the same effect for you that it does for football coaches. Setting a specific, ambitious goal and making a commitment to it will create a sense of urgency in you. It keeps expectations high for you. You're going to go for it. By setting this goal, you have committed—to both yourself and your students—to teach fundamentals to mastery and develop a deeper understanding of the principles and strategies they will need to ace every test and be able to apply those skills independently to solve problems on their own. How will you do that?

Both football coaches and teachers need to focus on results, so every decision you make as you begin planning for the year needs to support the specific goals you set to reach your vision. As we'll see, when planning backward to go forward, we start evaluating every lesson based on its effectiveness in getting results. If a lesson doesn't move your students toward that goal, you will have to consider discarding it or revising it. Are there lessons you really like teaching? Are there projects that have been part of your lesson plan each year for a number of years? Look at them critically and ask the question, "Will this lesson or project help my students be successful on test day?" If so, keep it in. If not, remove it from your lessons and replace it with things that will work.

When you begin the planning process by defining student success, you can see a clear path to success. Staying focused on what a successful student must do does two things. It takes the focus off the content and onto your students' achievements. Then lesson planning follows logically. If the results and expectations for students are spelled out from the beginning, planning becomes more focused. It becomes easier to plan what students need to know, understand, and apply to get results on test day. When faced with the variety of skill levels your students may have, you can coach them successfully to get results when you stay true to your goals.

Getting Started: What's Your Vision?

Even if you don't think so, you set goals for the year. Usually teaching goals are related to how much of the curriculum you will teach that year. Will you make it to Chapter 12? Will one class make it further than another? Maybe your personal goal is to get through the year with as few discipline issues as possible. These are survival goals. You strive to do as much as you can. Who could blame you if every student in your class didn't learn all the material you presented or some of your students got grades below a B?

What's your vision now? This isn't one of those in-service activities where the team gets together and comes up with a mission statement. Those are nice, but they rarely have the personal connection you need to move forward. They might be student centered, but they are rarely specific, measurable, or time limited. They have their place, but that's not what I'm asking you to do right now.

What is the commitment you want to make to your students this year? What will they be able to do by the end of the school year? What will you teach them? I decided that every student in my class would learn everything in my school's curriculum well enough to use those skills to solve problems independently on tests and practical problems during instruction.

Using the principles of coaching and backward design, or starting with the end in mind, I was able to get the students in my class to be confident math students. But the first step to success was developing my vision.

Now it's your turn. What's your vision for your students? Putting aside the roadblocks and barriers that you know you will face, what do you want your students to be able to do by the end of the year? What will your students know by the end of the year? Take the first step in making your vision real. Write it down here:

My Vision:

____

____

____

Setting a Goal to Reach Your Vision: How Will You Know When You Get Results?

A vision helps us put into words what we want to accomplish. It keeps us grounded. It lets us know what we want for our students. Now, you need goals to make that vision happen. The vision is what you want to accomplish in your classroom while the goals make the vision happen. Remember, goals need to be specific, measurable, time limited, and ambitious. I had three goals:

1. All of the students learn the whole curriculum.
2. All of the students receive a perfect score on every test and quiz.
3. All of the students leave the classroom at the end of the year as independent learners.

Your goals also should be measured often enough during the school year so you and your students both know if they are on track to get results. Football coaches use their visions to stay on track. Bill Walsh set up his playbook for each game, then set up practice plans so all his players would have the skills needed to make his vision happen. He included his players in his vision so that they could see the logic of the practice.

Be specific, check often

We give a lot of assessments to students in a year. Some aren't going to be helpful when you want to see results. You need data that are easy for your students to understand and that are given frequently and consistently to all your students. That's why using only midterm or finals aren't the best for setting goals. They only happen once or twice a year. That's not frequent enough to know if your students are mastering what you are teaching each day or for each test. State tests aren't helpful either. Finding out after the high-stakes assessments that your students haven't reached their goals won't help you monitor and adjust instruction throughout the year, because you receive the information too late. It may be helpful for planning next year but won't help you reach your goals now.

I followed these rules. My goal was specific and ambitious. One of my goals was for every student in my class to get 100 percent on every quiz and test throughout the year. Before setting this goal, I needed to consider all the possible ways I could tell if my students were successful. I wanted to know that how I was teaching them was working. For me, the answer was right there in my grade book. If I had decided I was going to go by what my students answered in class during instruction, I wouldn't have had a way to tell if everyone was making progress, because not everyone volunteers to answer. My goal wouldn't have been measurable in any consistent way either. I wouldn't have been able to prove to myself that my students were successful. By using the common assessments my school had developed, I was able to check everyone for understanding using the same assessments all math teachers were using in my school. Tests were given on a frequent basis throughout the year, which made them a good choice for measuring progress, both for my students and me. In addition to the common assessments, I gave quizzes frequently. I wanted to know my students were mastering each fundamental skill I was teaching them every step of the way. I wanted to know mastery was happening before I moved on to another concept.

Take a look at the assessments that you have available to you. Chances are you have a lot of them. Do you have common assessments ready to go? With the emphasis on Common Core curriculum, you may be in the process of developing them. Whatever you have available, pick the most specific assessments that are given often enough to help you stay on track. Weekly quizzes are better than portfolios due at the end of the semester. The remainder of this book will use test grades as the goal.

Make it about the kids, not the curriculum

Along with being specific, your goal should be student centered. It should be about what skills your students will learn. Are you going to look at fundamentals? How will you know when your students master them? Check your tests and quizzes to see if they fit the bill for your goal. What do you want your students to do once they learn those fundamental skills? Maybe you are ready to take your goal further to include application to real problems. Make problem-solving part of your goal by adding authentic assessments to your tests or quizzes.

Good coaches have plans to get every player ready for game day. They set high expectations for all. Just like your students aren't all math stars, Bill Walsh didn't have stars when he joined the 49ers. He made his players stars through his training program. Your goal will help you fill in the gaps for your students so they can get results on test day. For now, make sure your goal sets the standard for student skill and that you can frequently and consistently measure using the same assessments for everyone.

Make it ambitious

Make a goal that no one thinks you can reach. It isn't for anyone else but you. For me, it made it like a game. Could I make it happen? No one else needs to see it if you don't want to share it. But by setting this goal, you are taking the first step to focusing your planning and teaching so your students will make more progress than anyone else thinks is possible. Their previous teachers, their parents, and even their tutors probably wouldn't think your goal is reasonable if you told them anyway. That's okay. It's just for you, so just have fun figuring out how you will make it happen.

Don't be tempted to make a less ambitious goal. Go for it. Write it down and hold on to it. Then watch to see how your planning changes once you commit to it. It will guide you as you set priorities for what to teach. It will give you a focus when choosing lessons that further your goal, even if it means getting rid of some lessons you love teaching. It will also be ever-present in the back of your mind as you teach each day. See for yourself how it helps you get more out of your students than you have in the past. I'll be showing you how to reach your goal in the chapters that follow. Finding the time, setting priorities, and overcoming barriers to your success are all addressed. For now, commit to your ambitious goal.

Make sure your students will know when they get results too

Tests make a good choice for measuring your goal because your students are very familiar with them. They're used to looking at their tests to gauge how they are doing. Students need frequent feedback, and math tests are given frequently. Also, while quizzes are primarily to help you monitor and adjust instruction, they can add additional feedback for your students too. Seeing that they learned a tough skill or concept on a short quiz today is a good way to keep them motivated and coming to class energized tomorrow, because success can lead to higher motivation the next time you are teaching a tough concept. It's especially important to your students who have difficulty with math. Rather than just telling them they are learning more, you can show them by the improvement in their understanding of math reflected on their performance on tests and quizzes.

Your students will also be able to show their parents that they're doing well. Parents who worry about their child's achievement need to see results. A specific, measurable goal translated into test scores can do that, especially when you start to make some controversial decisions to save instructional time and take control of your students' learning by not assigning homework.

There are other ways to show your students they are math stars, which I will cover in future chapters too. Hands-on practice time in class also shows students they are successful. They can see for themselves by the activities you set up that they can do ordinary things extraordinarily well and are developing a deeper understanding of math principles. In later chapters, I will show you how to set up instruction so it includes time to work in class.

Keep Your Goal Handy: You'll Need It When You Start Planning

Congratulations! You have just taken the first step to getting results! If you completed this exercise, you have a vision for how you want your next school year to go. You know when your students will get results. And at this point, you may begin to realize that how you've taught in the past won't get the kind of results needed to reach your goal. Never fear. We have a plan.

Your newfound sense of urgency may begin to feel like panic right now. Sure, ambitious goals are great. But the stakes are high, and your students aren't all math stars. Do you even have enough time to reach your goal? How many days do you have to teach each concept or skill they will need? In the next chapter, I will show you how to start planning to get results by finding out exactly how much time you have to teach and help you get creative when finding more time.

Once you know how much time you can squeeze out of the year, you will probably notice that your schedule is going to be tight. In Chapter 6, I'll show you how to set priorities so the most critical skills will be mastered and can be applied to solve real problems while keeping expectations high.

Keep that goal front and center. Write it down. When the panic starts, I will show you how to coach yourself to stave off panic the way successful coaches do. When you face your students on that first day of school and each day after, it will help remind you what's important. Staying true to your vision will keep you on track. It won't do any good to rush to reach your goals if your students aren't on board, so I will also show you how building a team in your class will lead to growth in all your students.

Chapter 5

Yearlong Planning: How Many Days Do You Have to Cover the Material Your Students Need to Know...and How Can You Get More?

"First lay plans which will ensure victory, and then lead your army to battle..."
Sun Tzu, The Art of War

"Preparedness is the key to success and victory."
Douglas MacArthur, General, United States Army

"But I don't ever remember watching a [football] game on Thanksgiving and thinking, these guys aren't prepared. They simply have to figure out how to prepare in less time. It's a lesson we all could stand to learn."
Tony Dungy, Quiet Strength: The Principles, Practices, and Priorities of a Winning Life

Preparing to Win on Game Day

Have you ever run out of time to get your kids prepared for a test? What did you do? Put off the test and try again later? Send the part of the lesson you didn't get to home as homework? Have you ever gotten toward the end of the year and realized you weren't going to get through all the material to complete the curriculum? Have you ever had a great lesson your kids loved and you loved teaching, but their grades on the next test were disappointing? These things are frustrating enough, but the added pressure teachers receive to get students prepared and ready to do well on high-stakes assessments can cause panic. That is, unless you can harness that sense of urgency to work for you.

I knew when the order came from the top to not only teach the entire math curriculum but to also teach it for success that I would need to use what I learned as a football coach to develop a plan to get results for all my students. I knew I couldn't wait to see how the year unfolded. I had to be sure I would complete the curriculum by the end of the year. I couldn't take thirty days to teach one chapter, or I wouldn't get through all the material. I couldn't teach directly from the teacher's manual either. It didn't follow the curriculum I had to teach, and it took more weeks to finish than I had to get my kids ready. If I wanted to have my students prepared for each test, I needed to plan every day of the year. I had to plan to get results. I had to plan like Bill Walsh.

When Bill Walsh signed up to coach the 49ers, he had to rebuild an organization from the ground up. He had to take a demoralized team and a disorganized franchise and be ready to get results the first game. It was a big job, an overwhelming job. When he and his family went on vacation just before he started the biggest challenge of his career, it seems clear that Wash didn't relax. He never left their condo. According to his biography, when he emerged at the end of the week, he went home with a complete practice plan for his first season (Walsh,1990).

Bill Walsh was a planner. He had learned a lot about planning for each game when he was with the Cincinnati Bengals. His assignment as an assistant coach with the Bengals was to come up with a set of plays for game day. He learned from this experience that without a plan, you could get distracted by things that don't matter or panic and waste time rethinking what you decided to do. He planned more and more plays in sequence for each game day and won more and more games. His methodical way of playing football led to win after win, so it was logical for him to extend that idea to planning his practices. He planned and scrutinized every detail of practice, so no time was wasted. The way he rethought practice revolutionized the game.

Planning the School Year: Getting Prepared to Get Results

For high school coaches, football season starts right after the last season ends. The players are given a pat on the back for their good efforts and sent on their way. The seniors are wished good luck in college next year. In the quiet of the school halls, though, repairs are made, equipment is cleaned and set up for next year, and staffing for the weight room gets scheduled. And the coaching staff starts planning for next year.

Before setting goals for next season, coaches review of what worked and what didn't during the past season. Usually, this means listing the areas needing improvement. They might, for example, decide they need to be better on third downs. They watch film of the games to make decisions about plays and personnel. They take a look at their roster and make plans based on who is returning to the team. What are the strengths of their returning players? What do their players need to learn, and what do they need to practice to sharpen their skills so the team will get better results next year? The coaches decide on a strategy and develop plans to teach the specific skills needed to put that strategy in motion next season.

Bill Walsh took this preseason planning a step further. To make sure there was enough time during training camp to teach everything the players needed to learn, Walsh got out his training-camp calendar and counted the number of days they had available to practice. He took note when preseason games were scheduled. He considered the needs of the players during

The Science Behind the Success

Academic Engaged Time and Finding More Time to Teach

"Academic engaged time" is the time students spend on academic activities that lead to successful learning. It includes the amount and quality of instruction a student receives and their active participation in it (Gettinger & Ball, 2008). The more students are engaged, the more they are likely to learn. Dr. Maribeth Gettinger, a professor of school psychology at the University of Wisconsin, notes "that simply reclaiming the academic day (i.e., protecting and preserving scheduled time for core academic content) could nearly double the amount of instructional time students receive" (Gettinger & Ball, 2008). The first step, then, in maximizing academic engaged time is to figure out the amount of time you have available to teach math, then use that time to get the best results from your students.

practices. Players can't push hard every day. Lighter practices had to be included to reduce injuries and also to reduce cognitive load, the amount of information players can take in at one time. Before he could plan the first practice, he needed to know how many days he had for instruction to design a successful season.

Find as Many Instructional Days as Possible...

I knew from my years as a football coach that to be successful, you must plan. Success won't happen if you're not prepared. Goals aren't reached if you have no plan. The same is true in teaching. Students won't learn the math skills and strategies they need if you don't make every minute of instruction count. I knew I would need to be ready every day to teach my kids the skills and concepts they needed to win on test day. I sat down with my lesson plans and a calendar and got started. I asked myself three questions that day:

1. How many days did I have for instruction next year?
2. How could I get more?
3. How could I use those days wisely so every student in my class got results on test day?

Step 1: Calculate Available Days for Instruction Per Unit

My first concern was finding out the specific number of days I had available to teach next year. I took the 180 days on the school calendar and started to subtract the days I would not be able to teach. These included the first day of school (although I think you should teach math the first day!), the last day of school, track-and-field day, awards day, field trips, and two days lost to state testing. Together, this was seven days I couldn't teach—more than a week! I subtracted them from the 180 school days on the school calendar, leaving me with 173 days available for my lessons. That gave me 173 days to teach eleven chapters.

I knew my year would be tight, but I also wanted to be sure that my students had manageable assignments. That meant I had to look at the chapter tests. My school uses common assessments for math. All the students in my grade would be taking these same tests. As I reviewed the common assessments, I realized that one chapter was too long and needed to be broken into two units of study if my students were going to learn it well. That would reduce the cognitive load for them and help them master the material more easily, but it increased the number of tests to twelve for the year, which meant there were twelve units to master before the end of the year in 173 days.

Now I knew how many tests to prepare for and how many days I had to teach. Next, I needed to find out how many days I had to teach each unit. To start the process, I divided my 173 instructional days by twelve units and got about fourteen and a half days per unit. The realization was like someone dumping a cold bucket of water over me. About three weeks per unit? Could that be right? As teachers, we tend to feel we will always have the time we need to finish whatever instruction our students will need before the test. If a concept is tough, we take more time so we're sure our kids are ready, right? But as I looked at the available time to teach each unit, the reality was right in front of me. I couldn't take all the time I needed to prepare my students for test day. If I took seventeen days to finish a unit, I would need to pull three days from somewhere else. But from where?

The longer I taught, the more I became aware of how easily instructional time can be lost. I became frustrated about losing time to events that did not help students achieve academic success. When you lose a day, it's gone. You can't get it back. When we integrate lessons or have special days or events, we need to ask, "Are the students moving forward or sideways? What are we trying to teach them, and is it worth the loss of instruction?"

Then I started to speak up to save instruction days. I knew if results mattered, you need all the instructional time you can get. As both a teacher and a football coach, I learned advocating for what I needed and planning to use time effectively would get results. I needed more time to teach math, so I would get more time to teach math. That's where I would start.

...Then Find Some More!

Step 2: Get as Many Pure Teaching Days as You Can

My next job was to look for more time. I was working on a team with two other teachers in sixth grade that year. When I told them my plan to get all the material taught by the end of the year by keeping to my schedule, they were supportive. We put our heads together, and after reviewing our schedule, we found I could use some of the time during the two days put aside for state testing to teach math, because the full day wouldn't be used for testing. Now I had 175 teaching days—two more days to teach math to get results! Well, not quite.

I felt excited to get the two days until I realized that not all 175 days could be used for instruction. There were pretests and tests to give! I lobbied the assistant principal to remove the requirement to give pretests. They took up a full class period before the start of each unit, which I felt didn't justify the time I lost to instruction. I felt that, for the most part, they were a waste of time because they never gave me any information that I could use. There were always kids who did poorly on the pretests because they didn't take them seriously or they misread the question. A quick review was all they needed, and they were right back on track, but now they would be grouped with kids who needed more help than they did to understand the concepts I was teaching. The group was still made up of kids with very different skill levels, which required me to differentiate within the group anyway. Since I was the only one teaching math and I intended to set up my class so everyone could be successful using the principles of instruction I used in football, I wasn't going to need to teach the subject matter more than once anyway. If I was going to teach everyone as a team, I didn't need the pretests. The way I intended to set up my class and how I started the year with my students would make pretests unnecessary. I made my case to the assistant principal and got my days back. Getting rid of pretests saved me twelve teaching days for the year, enough to teach a short unit!

My lobbying efforts paid off. I gained two days of instruction from the state testing days. I rescued twelve days by not giving pretests. I still had test days to consider, though. I went back to my team and asked if we could all give our math tests on the same day during an extra language arts period that was built into the schedule. It would only happen twelve times in a year. There would still be time for extra language arts, so the team agreed again. Thanks to their support, I had 175 real instructional days to teach twelve units, or about fourteen or fifteen days per unit. Those days would be pure teaching days, not testing or pretesting. Fourteen or fifteen days to teach the skills, master the skills, and prepare the students for what they would face on test day was still a tight schedule, but I felt that I could make it work. I had squeezed out all the days in a year that I possibly could to get results for my kids.

Planning in this way was uncharted territory for me, but when I sat down and began to plan for the year, it was obvious that I needed to teach math this way. That sense of urgency I've talked about spurred me forward. But I also knew that some chapters would take less time than others for my students to master. Some concepts were just easier to understand, while others were always more difficult. I knew from experience that some units would take more than fifteen days for my kids to get results. That meant that if there was a unit I could do in ten days, for example, it would help me down the road when tougher skills required more days for my students to reach mastery. I needed to look at where I could pick up some more days for the tough concepts, and those easier chapters would help.

The more I worked on the plan, the more competitive I became. I talked to a fellow teacher who was also teaching sixth-grade math and told her about my plan. She told me it could not be done. In her opinion, there was no way to teach all the material in one school year, let alone plan how much time I would need for each unit and stick to it. Well, that was all I needed to hear to make me determined to prove that my plan would succeed. She had her chance. Now I was taking mine.

Planning Backward to Get Results Going Forward

Step 3: How Many Days Do You Need to Get Results for Each Test?

Now it was the time to plan out the year. Instead of starting with the lessons in the textbook and deciding how many days each would take to finish, I started with the tests I would be giving my students. In my case, that was the common assessments. Working back from those assessments, I then looked at the skills my students would need to take the test successfully. Then I decided how many days I would need to teach that material. Based on my experience teaching math, I came up with the number of days I would need to teach each chapter. For example, the first chapter, "Using Statistics to Analyze Data," was an easy chapter that I could teach in ten days. I subtracted ten days from my 175 teaching days and marked that the first test would occur, bringing me down to 165 days. But because I was below the fifteen days per chapter I had calculated as an average for each of the twelve units, I now had five days extra to use somewhere else. The second chapter, "Patterns and Algebraic Thinking," I reasoned I could do in fourteen days, but it would be ambitious. These were important concepts, and I would need to spend all those fourteen days to make sure my kids got results, but I didn't think I would need to use any of my extra days yet. I went through every chapter this way. I started with the average number of days and then decided if I needed more or less days. Some concepts I knew were harder to understand and master, so they got more days. When I finished, the plan looked like this:

Unit	Content	Number of Days Needed	Rationale
1	Using Statistics to Analyze Data	10	Allows 5 days to be used elsewhere
2	Patterns and Algebraic Thinking	14	Essential concepts so given the average time.
3	Adding and Subtracting Decimals	15	Ordinary concept needed to be learned to mastery.
4	Multiplying and Dividing Whole Numbers and Decimals	16	
5	Investigating Fractions	16	Basic concept needed to mastery
6A	Adding and Subtracting Fractions and Mixed Numbers	16	Too much cognitive load; split chapter
6B	Multiplying Mixed Numbers and Fractions	17	Very difficult concept to master, needed to take extra time
7	Ratios, Proportions, and Percents	16	
8	Tools of Geometry	15	
9	Geometry and Measurement	14	
10	Algebra	13	
11	Exploring Probability	13	Not taught to mastery; introduction so needed less time

We'll talk more about this in the next chapter, but this was the first step in using backward design. Instead of rushing through the book or the school's curriculum, I started with what my students would be expected to know. Then I planned how I would teach them the essential material to be successful on test day in the time I had available. Nothing was sacred. My favorite lessons or lessons that were a lot of fun had to help me reach my goal. If they weren't going to lead to success on test day, they had to go. My priority was to get results.

Coaching Yourself: Keeping the Goal Front and Center

Having taken the first step of my planning, I was excited to make my new plan work. I knew from coaching football to keep my goals in mind at all times if I wanted to reach them. Now I was putting the same principles into practice in my classroom. I would need to make sure my goals were in my face if I wanted to reach them, so I had my plan hanging up on the wall in my classroom and looked at it every day. Before this, I looked at curriculum guides but never thought about how many days it would take to teach each unit. Now my focus was on staying on track.

I also had a card on my desk with the message "You need to finish chapter __ by __(date)___." The date was already in my plan book, but I put it where I could see it all the time. It gave me the sense of urgency I needed to stay focused on the essential skills my students needed to learn for test day. I was going to get results by doing ordinary things, like teaching and planning, extraordinarily well.

I had another reason to keep my goals front and center. I had decided to stop giving homework. It was a big decision, one that I talk about in detail in Chapter 10. When I thought about it and looked at how many precious instructional minutes I was wasting in class every day managing homework, I decided there were more reasons *not* to give homework than to give it. So, I planned for that and adjusted my teaching accordingly. I had to be sure my students mastered the material I taught them every day. This ensured they would be alert and engaged with me, but it did increase my sense of urgency and put more pressure on me to get results for my students every day and on every test day.

Getting Started: How Much Time Do You Have?

Now it's your turn. Can you get all the material your students need to learn in the actual time you have to teach? In three simple steps, you will have taken a giant leap toward getting results on test day. Grab your school calendar and the worksheet below. It's time to calculate how many days you have available to teach math next year and calculate available days for instruction per unit.

1. *How Many Days of School Are There?*

You probably already know how many days are on your school year. (Who doesn't?) Put that on the worksheet below.

2. *Subtract Days Lost to Instruction.*

Now go through that calendar and count the number of days you lose to instruction through the school year. These are any events that take students out of your classroom during times you would be teaching math. Start with the last day of school, track-and-field days, awards days, field trips, high-stakes testing days, and in-service days. Don't forget to include special events your school may be doing or do on a regular basis, such as project presentations. You'll notice that the first day of school isn't on this list. I feel you should be teaching that day. Try to be as accurate as possible at this point. As our coaches like to point out, the more prepared you are, the less likely you are to panic. Subtract the number of days you found you can't teach math from the number of school days per year to get your total of true instructional days.

It is a good idea to make sure that these really are all instructional days. Be sure there aren't other regularly scheduled activities that will take time away. For example, I had pretests to contend with. You may also have scheduled activities within your class that take time away from instruction. Be willing to advocate for instructional time if you need more time to teach. I showed you how I found extra instructional time in my situation. Take a look at your situation and get creative about finding more days to teach.

3. *How Many Tests Will You Be Giving and What's the Average Number of Days You Have Per Test to Get Results?*

The next step is to begin planning the year by looking at the tests that your students will be facing. I had common assessments that my team had already done, so I was ready to look at the units of instruction that applied to those tests to decide how much time I would need to teach the concepts my students needed to be successful on the tests. What tests does your team use? If you are using Common Core, you probably are working on your common assessments or have them done. You can start planning backward from those tests to determine the skills and concepts your students need to get results on test day. In future chapters, we'll start to build the lessons that will make this happen and set up a classroom environment where most of your students will get results. For now, calculate the number of days you have for each chapter by dividing the number of true instructional days you have for the year by the number of tests you're going to give to get the average number of days you have for each test and chapter.

Number of Days in School Year:	______
Subtract Number of Days on the Calendar Lost to Instruction: Such as last day of school, state-testing days, in-service days, etc.	− ______
Subtotal of True Instructional Days:	= ______
Subtract Days Lost to Instruction within Total Instructional Days: Such as days used for testing, etc.	− ______
Divide by Units of Instruction	÷ ______
Average Days Available for Instruction per Unit	= ______

4. *How Many Days Do You Need to Get Results for Each Test?*

Now it's time to figure out the number of days you need to teach the skills needed for each test. If you've taught math for a while, this should be fairly easy. You know if concepts tend to come easily to your students or if they are hard to understand and may take more time to master. Be sure to consider the cognitive load of your lessons. As a quick rule of thumb, the newer the concepts included in a lesson, the heavier the cognitive load. If you're not sure how many days it will take you to teach each lesson, plan for the average number of days per unit and use that to determine how many days you need. When you save a few days, you can use them for more difficult chapters down the road. Keep track of this to help you plan for next year. You can use the table below to help you work out the number of days you need for each unit, then take note of the number you needed, which will help your planning next year.

Chapter	Content	Number of Days Needed	Notes for Next Year

Getting Ready for the Next Step

So far you have set goals that no one thinks you can reach and let a sense of urgency develop in you to spur you on so all your students will get results on test day. Now you have the first step of your plan to make sure you make that goal. You have begun to plan backward by calculating the number of real instructional days you will have next year. That sense of urgency to reach your goal probably just kicked up a notch. You might have also realized that with so few days to teach tough concepts, your instruction must be focused so it can be done in the shortest time possible and still get all your students to mastery. You must make sure all your lessons are using the most effective instructional strategies as possible and lead to real improvement in your students' skills. Nothing is sacred. You may have the best lessons, but if they don't lead to success on test day, they have to go. Basic skills have to be taught to mastery so your students can apply them to tasks that require higher-level thinking skills, like problem-solving. In the next chapter, I'll show you how I prioritized the material that needed to be taught to mastery so my students would get results in the number of instructional days I had available. Using backward design, I developed lesson plans that teach ordinary things extraordinarily well as well as build a deeper understanding of math that my students would carry with them to test day and be able to build on for future lessons.

Chapter 6

Prioritizing Lessons: What Do Your Students Need to Know to Get Results?

"When you fail to plan, you are planning to fail."
Benjamin Franklin, American Statesman

"We only practice what we do in games."
John Gagliardi

Picture this: you just got plowed by the other team's lineman. You're flat on your back trying to catch your breath, staring up at the sky.

It appears to have been a powerful learning experience for John Gagliardi. When the coach of his high school football team left to fight in World War II, John Gagliardi, a junior and sixteen years old, talked the parish priest of his high school into keeping their football team together and, in the process, got his first coaching job. He became the team's new coach and the quarterback.

Football practice back in the 1940s was all about calisthenics. The thinking was if the guys were fit, they could get results. That's fine if you're the coach on the sidelines, but John noticed if his team made a mistake or the opposing team changed their strategy, his players didn't know what to do next. The traditional practice didn't prevent mistakes on the field or help his teammates understand their roles in plays. Unless he wanted to get used to being creamed every week, he was going to have to rethink the way he ran practice. It didn't take him long to change from calisthenics to frequent repetition of fundamentals practiced within the context of the plays they would be running that week. It worked. His team started winning. (Weeres, 2010)

For the rest of his career, Gagliardi continued to place a strong emphasis on high accuracy for fundamentals and a deep understanding of the role of each player in their plays. It wasn't enough for players to learn skills. They also had to answer the question, "What would you do if?" He prioritized teaching the fundamentals so no mistakes were made on game day and making sure everyone understood the strategy behind every play they would run.

Like football, math is a game of mistakes. Fundamentals must be mastered and executed flawlessly to win on test day. But it's not enough to know fundamental skills. Students also must know when to apply them. That takes a deeper understanding of math concepts and the role those fundamental skills play in solving problems. When faced with a tidal wave of facts, concepts, strategies, and standards, how do we develop sensible plans that help our students become math stars?

Planning is key to all our coaches' styles. As we learned in Chapter 5, Bill Walsh started with the number of days of practice he had, so he knew how much time he had to make it all happen. With their jobs on the line, our coaches left nothing to chance. Their plans always included both mastery of fundamentals and that deeper understanding of their system so their players could see the big picture, and, in some cases, help develop the strategy for game day. John Gagliardi's quarterbacks could run plays themselves and went into the game on game day knowing that they had the final say for decisions on the field. Bill Walsh set up his playbook long before the first game, so he knew what skills his players needed to master to execute them flawlessly on game day. John Gagliardi expected his players to go beyond and understand not just what they had to do on the play but also where they fit in the team's strategy for winning.

Glen Mason and Bill Walsh both took teams that had fallen apart, and each looked at the needs of their team and its members to decide how to get the franchise back on its feet and winning again. Glen Mason started out with a thorough evaluation of the state of his new program, or a needs assessment, of his team and found out what worked and what didn't. He didn't dwell on the problems. Instead, he developed a plan to fix those problems one by one (Mason, 1995). And Dennis Creehan didn't turn around the many teams he coached by just hoping for the best. He set goals for every practice and developed a teaching strategy so fundamentals would be mastered by providing as many reps as possible (Creehan, 2000). All these coaches had detailed plans to teach their players fundamental skills and strategy based on that plan for every practice.

In this chapter, we will build on the goal setting discussed in Chapter 4 by beginning the process of setting priorities for lessons that lead to confident, competent math students. We want to build that confidence by teaching fundamental skills to mastery, the focus of this chapter, while also teaching a deeper understanding of concepts so students can apply them on their own on test day. After completing the exercise in Chapter 5, we know how many days we have to prepare for each test day. We've squeezed as many days out of the school year as we can, but the schedule is still tight. Now it's time to develop our plan so we can teach the curriculum while keeping expectations high, by planning backward to go forward.

Start to Develop the Plan, Teach the Plan, and Keep Expectations High by Planning Backward

In my first years as a football coach, I needed to plan out the year. What was my goal? Win every game. No one expected me to win every game. That didn't mean that I shouldn't go for it. I was keeping expectations high, not just for my players but for myself. It focused my planning and gave me the sense of urgency I needed. What did I want the players to achieve? I wanted to have the players ready for the game. "Being ready" meant doing fundamentals accurately every time and understanding our system and their part in it. The first step in preparing them for game day, then, was to teach the fundamental skills they needed to run our offense, defense, and special teams.

> ***The Science Behind the Success***
>
> **Motivated to Understand**
>
> Learners, according to cognitive psychologists, are motivated by the "search for meaning" (Bruning, et.al., 2011). Providing a deeper understanding feeds that need by providing ways to incorporate what they already know with new information.

There are fundamental skills that each player must know in order to play well. After I was done teaching them those skills, they would know how to perform them extraordinarily well, or to mastery. Those skills had to be second nature to my players so they could perform them consistently regardless of the pressure they might feel. If they couldn't do that, the play wouldn't be successful on game day. But that wasn't enough. Not only did I want them to have those fundamental skills needed to play well, but I also wanted them to have a deep understanding of our system. I wanted them to know *why* they were doing each skill and play. I would not be satisfied with their knowledge until they could apply our system on the field during a game with confidence and get results.

I didn't know it at the time, but I was using backward design. Backward design starts with desired results. People in many professions have been using this process for years. Business leaders are taught to "start with the end in mind" by Stephen R. Covey in his very famous book, *The Seven Habits of Highly Effective People* (Covey,1989). Highly effective football coaches set clear goals for their teams each year, long before the next football season starts, with detailed plans for teaching them. In their book, *Integrating Differentiated Instruction and Understanding by Design (UbD)*, Carol Ann Tomlinson and Jay McTighe describe how to use this concept to help teachers plan effectively. By first identifying desired results, then determining acceptable evidence, instruction stays focused on student skills when teachers plan learning experiences. By identifying not only the skills needed to do well but also the big ideas that lead to a deeper understanding when using those skills, teachers ensure that students, like my football players, learn how the skills and actions they are performing get results.

Planning Backward to Get Results

There are so many expectations and demands for teachers that it's easy for the big picture to get lost. Tomlinson and McTighe talk about how those forces can result in disconnected lessons that lose the big picture. They believe the goal of education is to provide the understanding students need to evaluate information for themselves while still meeting the standards and goals of your school.

To accomplish this, they advise planning backward when developing lessons for the year. Most times, planning begins with the first lesson and proceeds in a logical order to the end of the curriculum. But how do we know if we will get results? Will our students come out at the end of the school year competent, confident learners? When beginning the planning process by defining student success, we can see a clear path to results. If the results and expectations for students are spelled out from the beginning, planning becomes more focused. Knowing what's expected, it becomes easier to plan what students need to know, understand, and apply. That's your first step. There are skills that must be mastered so they can be done automatically or very efficiently. They then need to be applied to basic strategies that are used to solve problems. Setting priorities for instruction involves deciding which of these strategies need to be learned and used throughout the year.

The second step in planning backward is to "determine acceptable evidence." Tomlinson and McTighe advise teachers to consider in advance what evidence they will use to decide when students are successful based on the desired outcomes for the subject. In Chapter 4, we set our goals based on our vision for student success. The evidence is needed to show you made your goals. It could be your classroom work, quizzes, or common assessments, for example. Keeping in mind what students will be expected to know and how they will be expected to show it helps focus lesson planning as well as teaching.

Now it's time to plan the activities and learning sequences and allocate the available resources needed to reach your goals. Tomlinson and McTighe believe teachers need to stay mindful of the outcomes of each lesson and ensure it will result in necessary learning. Many activities in middle school can take a great deal of time and be fun or interesting, but as we saw in Chapter 5, there are only so many days to teach in a school year, and that time must be protected. If a lesson doesn't lead to skills mastery or big-picture understanding for your students within the time you have, consider finding a more effective instructional technique. Lessons need to take students forward, not sideways.

My goal was for all my students to get an A or a B on the common assessments, so I started by looking at those tests to decide what fundamental skills my students needed to know on test day and the strategies they needed to have to apply them in the right situation. Tomlinson and McTighe urge teachers to keep the goals in mind when planning activities, staying mindful of the outcomes of each lesson to ensure it will result in necessary learning. I found that many math activities are fun and engaging but can take a great deal of time. If it didn't lead to success within the time I had, I had to find another way. The activities used to teach skills and the understanding to apply them must be aligned with the goals you set. All objectives and activities that are included in your plan must help to get results on test day.

I believe the techniques I learned in *UbD* can help all teachers plan and teach for true student understanding. This model is much too comprehensive for us to cover here, but I encourage you to take a look at their book. Backward design helped me the most when planning lessons for the fundamental skills my students needed to make fewer mistakes on test day and provided the foundation to acquire the knowledge they needed to gain a deeper understanding of the math concepts I was teaching. Even before I used *UbD* as a math teacher, however, I realized I had been using it as a football coach.

Using UBD to Teach the Wing-T

Let me show you what I mean about using backward design by planning out a Delaware Wing-T, called a 121. This is a wing-right, sweep-right play where we will pull both guards, fake a fullback up the middle, waggle the quarterback to the left, and run the halfback on a sweep to the right. To explain what each player needs to do would fill up many pages, so let's just work on the guards for this example. Guards are part of the offensive line. Their job is to protect the quarterback and help running backs break away.

Step 1 is defining our desired results. At the end of instruction, we want our guards to know key terms like "pull," "gut," and "trap." We want them to be able to block their assigned opponent so that they do not make the tackle. These are the fundamentals. All guards need to do these fundamental skills to mastery for the plays to work consistently. We also want our guards to have a deeper understanding of our system. They need to know why the play works and their role in running the play successfully. When they do their part correctly, the other members of their team can do their parts successfully too. In this case, we want the guards to understand the techniques they need to complete to make the play work under varying circumstances. John Gagliardi did this with what-if questions. The essential what-if questions for our guards in this play would be: What do I do if my player runs upfield? If they crash hard? If they just sit there? What do I do for different defenses? What if nobody shows up for me to block?

Step 2 is to determine acceptable evidence. In other words, what are the assessments I will use? What performance tasks will show evidence that the guards can run play 121 successfully? If the player can block their opponent so the opponent can't make the tackle, that shows evidence that they have performed the task at the desired level. We will make this evaluation in practice and again after watching the film of the game.

Step 3 is the learning plan. The learning activities we planned started with the bird dog drill. This drill teaches the players how to take their first step with their foot. The guards would come out of their stances, stay low, and take the first step for each play we ran. When that was mastered, we worked on shoulder skills. This helped players learn what to do as they made contact with the opposing player. Next, we worked under a "chute," hitting a bag to mimic coming off the ball and making contact in a game. Our next drill was to place the players in the same position they would be in for our offensive formation and run the guards together just like they would in a game. The guards used everything they had just learned plus additional instruction on footwork and position. Later, the guards worked with the other lineman as a group and then with the whole team. Learning both fundamentals and their role in the play made them better able to help their teammates get results. It wouldn't matter if they understood how the play helps the team if they don't know how to take the first step. It wouldn't matter if they knew how to block if they didn't know when to do it. Incomplete knowledge won't get results on game day.

Using UBD to Teach Adding and Subtracting Fractions with Unlike Denominators

I was already familiar with backward design from my football experiences and knew how it could help planning for success. I wanted all my students to go into every test and quiz with confidence, knowing they could be successful. I wanted to prepare them by teaching the skills needed to be successful in all math situations they would face on test day. I needed to break down the skills for each chapter and decide how to teach them so my students would learn these skills and concepts successfully within the time I had. In math, there are skills that must be mastered so they can be done automatically. They then need to be applied in basic strategies that are used to solve problems. Setting priorities involved deciding which of these strategies needed to be learned for the next test and mastered so well my students could use them when needed throughout the rest of the year.

Let's look at a math example. Planning backward to teach students to add and subtract fractions with unlike denominators starts with the final goal: students will be able to add and subtract fractions with unlike denominators accurately. When students have mastered this objective, they should be able to work a fraction problem, during which they will add or subtract fractions when both fractions need to be changed into least common multiples (LCM) and then put the answer in simplest form. Working backward from there, I worked on problems with LCM, but my students would only need to change one fraction and put the answer in simplest form. Also, going backward, they would need the skill of equivalent fractions. And to understand equivalent fractions, they would need to understand LCM and least common denominator (LCD). If simplest form is a skill needed (changing improper fractions to mixed numbers and reducing fractions), then I would need to teach it before the previous step. Going backward from there, I could see the first problem in the unit needed to be adding and subtracting fractions with like denominators where the answer was already in simplest form. Let's see what it will look like from the beginning to the end of the unit with the skills students need to master to have the confidence to do well on the test.

Step 1 is defining our desired results. Students should know how to add and subtract fractions. They should understand that denominators must be the same before you can add them together. They should be able to add or subtract fractions that have unlike denominators. They should also be able to change both denominators and always put their answer in simplest form. A deep understanding of LCD will help students achieve success in this unit of study.

Step 2 is to determine acceptable evidence. Each student needs to show mastery of each skill leading up to adding and subtracting fractions with unlike denominators. First, students will be able to add and subtract fractions with like denominators. Then, students will add and subtract fractions and put their answer in simplest form. Lastly, students will change both fractions to like denominators, add or subtract, then put their answer in simplest form.

Step 3 is the learning plan.

1. Add and subtract fractions, like denominators, answer already in simplest form.
2. Teach simplest form.
3. Add and subtract fractions with like denominators, change answer to simplest form.
4. Teach equivalent fractions, LCM, LCD.
5. Change one fraction, add and subtract, put answer in simplest form.
6. Change both fractions, add and subtract, put answer in simplest form.

If I wanted to put my students in a position to win, I needed to be sure the most important skills they needed for the next test were on my list of priorities. Observing my players let me know if the group was mastering the skills or if any individual players needed corrective feedback during practice. Observing my students wrestle with the lessons I was teaching let me know when I needed to break down more complex skills into each sequential step or drop back and teach prerequisite skills first. On the other hand, if they were flying through the lesson, I could move on and bank the time for lessons they may struggle with in the future. I may have allotted twenty minutes, but if they got it in ten, I could move on. For example, when teaching adding and subtracting with like denominators already in simplest form, it was likely going to go quickly. After teaching, doing some reps, and providing corrective feedback, I gave about five sample problems, and if they answered them correctly, I was confident they understood. I was ready to move on to the next skill in the sequence.

Getting Started

Step 1: Defining Our Desired Results. What is my goal?
Football coaches decide what they want their team to achieve and then start to think about what skills and knowledge they will need to have to reach that goal. The goal itself is not difficult for them to come up with: win every game. The best coaches, though, go a step further. They want their players to master basic skills so they can do ordinary things extraordinarily well and know their system well enough to make decisions on the field, increasing their chances of winning on game day. My goals were to complete the objectives in the math curriculum so my students mastered basic skills and would be able to apply them to solve problems independently. But as we learned from *Understanding by Design*, the goal must be testable. You need to be able to know when you reach it and prove it to others.

Step 2: Determine Acceptable Evidence. How Will I Know When My Students Have Reached the Goal?

If you set your goal in Chapter 4, you're done with this step. If not, it may be time to go back and go through the decision-making process we did in that chapter. For example, think about the tests you give your students throughout the year. Which are given the most frequently? Are your team's common assessments linked to the high-stakes test? They can be your markers along the way in the school year to see if all your students are learning. Do you give a practice test as a benchmark during the year? These are possible choices to consider. Developing quizzes based on each step of the process can tell you if students are staying with you. I used them after each skill to check for understanding. More on that later.

Step 3: Develop Your Learning Plan. Set Priorities for Mastery of Fundamentals and Successful Application of Skills.

Tomlinson and McTighe (2006) suggest you look at the big ideas that students can take with them each time they approach a similar problem. Rather than teach facts, they advocate, teach strategies for learning along with the facts. Football coaches would agree. They consider the skills their players have and develop a lesson plan that teaches the skills their players may lack to be successful against the other teams they will meet any given game day. I set my priorities based on what was expected on the common assessments, then planned backward, making sure that the prerequisite skills needed to understand the concepts were part of my lesson plans.

Set your priorities the same way. What do your students need to master to efficiently apply basic skills to the problems they will face on each test? These are the skills that require frequent repetition until they become automatic. What are the types of applications they will need to understand to successfully solve problems on test day? These are the big ideas they need to know how to tackle. Real-life problems help students remember when to apply these strategies. Activities to ensure your students can apply skills, or authentic assessments, should also be planned and done under your watchful eye in class. More on that later.

Keep Your Plan Handy. It's Time to Set Up Your Class.

When Bill Walsh walked on the field for the first game with the 49ers, he had a set of plays already set based on what was needed to win the game. He put together a plan to win that rivaled most military campaigns. But long before the first game, even before the first practice, he had a plan to teach each skill each position needed to know and what each player needed to master to win on game day. Those skills were chosen carefully to be done flawlessly every time. Anything less could cost the team the game. Even more important, the skills that were taught were the ones needed to deliver the strategy he had developed for his team that year. Long before that first game, his vision for his team was set, the lesson plans for his practices were planned, and every player on his team learned his system. When he walked on the field on game day, he and every member of his team were ready to get results. Being a master at planning made his team a winner.

Starting with the end in mind helped me to see the plays that were needed and the skills each of my players needed to be ready on game day. Those fundamentals could be built into plays that set up strategy to win against that week's opponent on game day. By breaking down the key concepts that my players needed to know to understand the play, I helped them learn each step in the plays we would be doing on game day. As they made each move to complete their part, they knew they were ready. They had practiced each step over and over until it was second nature. Because they started with the end of the play and learned backward, they knew where they were going right from the beginning. That was a good start, but it wouldn't help if they didn't know when to use it or what to do if the situation in front of them didn't come out like we planned. Asking the question, "What would you do if?" like John Gagliardi or practicing Bill Walsh's situation football, described in the next chapter, would help my players make judgments on the field to complete their role in the play successfully.

When I started planning the same way for my math students, they started to get results that I could see each test day. They were prepared to do fundamental skills without mistakes because each step they needed to take had been broken down to include the essential ones for the test. If they didn't know the precursor skills, I put that in the plan too. Some of those skills I knew they would need from experience with previous classes. Others were included in the plan in case I needed them. I wasn't going to lower the expectations for some students who might not know math very well coming into my class. Everyone was going to be successful.

Setting my goal gave me a sense of urgency, but my planning kept me from panicking when I saw how much I needed to teach in the time I had to teach it. Setting priorities and looking at my lessons with a critical eye kept me from wasting my valuable lesson days with activities that didn't move my students forward. Keeping my goal and my playbook close kept me from drifting off my plan or trying to wing it if things didn't seem to be going well in class one day. With the fundamentals in place, my next step was to make sure my students knew in what situations to use them and got plenty of chances to apply them.

Practice:

Teaching Your Students to Be Math Stars

Chapter 7

Setting Up Your Playbook to Get Results

"No aspect of coaching is more important than teaching."
Bill Walsh, Finding the Winning Edge

"The goal of planning is to remove from a tough situation the panic element of 'what the hell are we going to do now?' The less thinking people have to do under adverse circumstances, the better. When you're under pressure, the mind can play tricks on you. It's a terrible mistake to let outside forces influence you more than the pragmatic realities of the situation already are. Better to reduce the effect of outside variables and revert to something you've practiced and practiced. When things get tense in a football game, I want to be able to make decisions clinically, because I've thought them through beforehand, rather than take an ad-libbed, seat-of-the-pants approach."
Bill Walsh, Finding the Winning Edge

Setting up your playbook for each lesson each day is the last step in planning to get results. After this, it will be time to step into your classroom and teach math so your students will learn each skill they will need for every situation they will face on test day. This goes beyond typical planning to planning each minute of each day. It's the planning that will help you reach your goal for each of your students to get an A or B on every test and build confidence in themselves while they do it.

Most coaches plan for how plays will be taught by working backward like *Understanding By Design* taught. They first decide what plays they would like to run for the next year. The list is usually too big, so they have to cut back. Once they have their playlist, say for offense, the skills are broken down for the quarterback, running backs, receivers, and linemen. Each player will have to master the skills needed to run each play effectively. All areas of the game are covered. The players must be prepared to handle all situations that may come up in the game. Players need to perform the play correctly, consistently, and at the right time.

I'd set up my units for the year using backward design, setting priorities for which fundamental skills needed to be taught to mastery and making sure to include the big picture for applying those skills. I needed to plan those lessons to keep expectations high because my goal was all my students were going to learn the material they would need to get results on test day. There would be no exceptions. And they were going to learn it from me, nobody else. To make that happen, I needed to set up my lessons each day to use the time well, keep the attention of my students, and make sure everyone was learning each skill or strategy well enough so they could apply it on their own on test day. How was I going to make those decisions, keeping my goals in mind? Two coaches I'd read about came immediately to mind.

Situation Football in a Racehorse Classroom

As we discussed in Chapter 3, all successful coaches implement their plan so their players can learn the skills they need for game day as quickly and efficiently as possible. They don't have a choice. There are only so many days to practice, and the team must be ready in that time. The more effective they are at this, the more likely they are to reach their goal to win every game. While all our coaches are very effective at doing this, there are two that stand out as exemplary. Bill Walsh went so far as to write a handbook for coaches and included the steps he felt were needed in the teaching process to be sure all players learned what they needed to get results. The other coach that took great pains to make every minute count in his practices is Dennis Creehan (Creehan, 2000). His racehorse practice keeps things moving, so there is time for instruction, practice, and repetition of fundamental skills without overwhelming players with too much information at one time while providing time to practice under the watchful eye of the coach.

In his book, *Finding the Winning Edge*, Bill Walsh makes it clear to prospective coaches that teaching players fundamentals is critical to the success of the team. It is the coach's most important responsibility, as far as he was concerned. The team isn't going to win football games unless the coach knows how to teach fundamentals to mastery. Properly executed fundamentals are the foundation of a good football team.

There are a lot of things that can go wrong on game day. Walsh feels you need to prepare for anything that might happen on the field, so you aren't taken by surprise. With that on his plate, the last thing he wanted to worry about was sloppy execution of the fundamentals of the game. Teaching players needs to include clear, concise instruction with consideration of the needs of the learner, demonstration of the skills, hands-on experience under the supervision of the coach, and timely corrective feedback until the players master the skill. His practice plans were very detailed and put together ahead of time, so no time was wasted during practice. He left nothing to chance.

In addition, Walsh feels that the better the players understand the game of football, the more likely they would win. To get results, he feels that players need an overview or context for the skills he would teach as well as mastery of those skills. Players need to see how the material relates to them and know from the presentation that it is important to the coach that they understand it. He feels coaches also need to be aware of the individual differences in skills and experience that their players bring to the lesson and adjust the presentation to these individual differences. Even after he rebuilt the 49ers, this priority did not change. His lessons explained material clearly in a way that everyone, regardless of their experience, would understand, and he kept everyone's attention while they did it. He feels coaches need to have a logical, easy-to-follow presentation. A sequential approach to teaching a skill allows the players to follow the building blocks to successful mastery.

Time is always a consideration for football coaches. They don't have all the time they want to make sure these fundamentals are learned to mastery. They have only so many practices in preseason and only so many hours they can push players once the season starts. They need to use that time efficiently by choosing effective teaching strategies to get results.

What Walsh calls "situation football" not only considers the skills the players need but also their application to real situations that players would find themselves in during games. Plans are developed based on the strengths of the current team and the best way to use their talents against their opponents. Walsh had to look at the situations that they were preparing for. The strategies and situations that would come up in every game were broken down and added to the plans for practice. How you play the game at the kickoff is very different from what you do when you are ten yards to the goal. Walsh analyzed and prioritized all situations based on how many times they would need to be practiced in order to get results during a game. Then those plays were put into the schedule.

Like Bill Walsh, one of the problems I faced in my math class was getting enough practice time in to master basic skills for the situations my students would face on test day in the time I had. My schedule was tight, and I had no time to waste. How could I keep my students on track and give them enough reps to reach mastery? I decided I would do it the racehorse way. Coach Dennis Creehan dubbed his practice technique with the term "racehorse." The goal of a racehorse practice is to get the most repetitions per practice as possible. During a racehorse practice, everyone works with speed and a sense of urgency. For each player to get enough reps to learn fundamental skills, coaches must have a clear understanding of the goal for practice before they even walk onto the practice field. Like Walsh, planning the day's practice is very important to this process.

The first step in running a racehorse football practice is a meeting of the coaches. All new concepts are put on the board, and the coaching staff develops the plan for each practice. All decisions about what will be covered in practice are decided then, along with the amount of time allotted to master each skill. Skills are prioritized, and the minutes of practices are assigned based on their difficulty or importance to getting results on game day.

Setting Up My Playbook for Situation Math in a Racehorse Classroom

As a football coach, I had a plan for every minute of practice. My time was tight, and I wasn't going to get any more. There were only two goals for practice: to improve skills and to prepare for the next opponent. These goals had to guide my planning. There were a lot of fun or interesting things we could do at practice, but if they didn't help my players master fundamental skills or get ready for their next opponent, they weren't worth the time. I needed to allocate the time for practice wisely. Priorities for each practice were set based on what my players needed to learn to be prepared for the situations they would face on game day.

As coaches, we looked at films of the other team and pinpointed our opponent's best plays. We needed to use our limited practice time to prepare for that team's best plays, the ones that would be toughest for our team to beat. We never wanted to put our players in a situation they had never seen before. Luckily, there are a number of plays that have similar features. If our players could recognize those same features in other plays against other opponents, they would be better prepared to beat those plays too. By showing our players how to deal with that situation, not just a specific play, they were more likely to be successful on game day.

If I wanted to run a situation math class, I couldn't just put up a problem in class and explain how to get the right answer for just that problem, because they would never see it again anyway. I wanted my students to learn to address situations that they might find with any problem of that type. So, I wanted to work on situations that they might face on test day. In football, you may work on goal line plays the other team may run. In math, you may work on how to do LCD. For example, I

The Science Behind the Success

Success Increases Engagement

Success is a large part of your students' active engagement. The more students actively participate in learning activities the more engaged they are and the more likely they are to be successful. They are more likely to be actively engaged in your lessons if they have a high rate of success on challenging but reachable tasks. (Gettinger & Ball, 2008).

asked my students, "What would you do if you finish that problem and it's not in simplest form?" If they learned to recognize what to do whenever that situation came up, they would be better prepared to deal with the situations they would face and get better scores on test day.

Another of my goals was to build confident, independent learners. By starting with situations and teaching my kids to recognize them, I was making it possible for them to have strategies for dealing with similar problems, or a big picture for math. Going forward in my plans, I wanted the students to recognize similar situations and know what to do. The more prepared they were for what was coming on test day, the better their grades would be. The better their grades, the more confidence they would have in themselves as math students. That confidence would help them face the tough lessons coming up, knowing they could be successful.

I knew that I couldn't spend the same amount of time on all the things my players might see on game day. As a math teacher, I couldn't spend the same amount of time on all concepts. I had to prioritize to fit my tight timelines.

As a football coach who sets those priorities in preparation for facing our new opponent on game day, I asked myself, "What's the toughest play the other team has? What's the one that is going to trip us up?" Once I knew what that play was, I asked, "How can we take away their best play?" I let my players know that we were not going to let the other team use that play. Then we set up a strategy to defeat their best play and set it as a priority for practice that week.

For my math class, I started with the opposition: the test. Some problems would not pose too much difficulty for the students, based on their performance on past tests and from my experience teaching the material in previous years, but I wanted to find the situations that would most likely trip them up. The situations that my students would need to beat on test day set the priorities for what I would spend my time on that week.

If I wanted my students to learn the techniques they needed to beat that tough problem and build the confidence that comes from mastering tough situations, breaking those techniques down to an easily understandable level was key. I wanted everyone to end each step of the lesson successfully. As a coach, I looked at what I wanted my players to learn and then decided how much it needed to be broken down to ensure mastery. On the practice field, there were some techniques I knew my players could not learn to mastery without breaking them down into several parts. When teaching blocking skills, for example, I started with the end in mind by putting the players in a fit position, the last step in blocking, leaning into the bag in the perfect position. Once my players knew how they were supposed to end up when they hit the bag, I had them go back on all fours and hit it, then stand up and hit it, then go back to where they were supposed to run from and run in and hit it again. Then, when we put the whole play together, they knew the proper techniques and where they fit in the play. I did the same thing in my math class. When I was teaching, I also had to break down each technique as far as necessary. I wanted to be confident that my class could master the skills and techniques they needed to to be successful facing the tough problems they would see on test day. While I didn't want to waste time teaching them what they already knew, I couldn't rush them through something they hadn't mastered yet. By breaking down skills and having lessons ready to teach if I needed them, I was ready for whatever my students might need.

Now that my priorities for the week were set, I set my goals for each day's lessons and made my plans. Those plans were done to the minute. In football, I might use seventeen minutes for one skill but only five for another skill, based on my priorities for practice. Those priorities were based on the skills my players needed to prepare for the situations they would face and increase their chances to win on game day. I needed to do the same in my math class. I could have just planned my lessons based on the curriculum and spent the same amount of time on each concept or skill, but then a day would be wasted going over something the kids already knew and not enough time would be available for the tough concepts. My lessons for the week needed to make the most of the time I had, so I gave the most time to the situations my students needed the most practice to master.

Football coaches are very regimented in how they run practice. Every minute is planned out. It is not uncommon to spend fifteen minutes on group offense, for example. Then a horn or a whistle blows, and the players run to the next thing, trying not to waste any time. It's all written down on paper and followed to the second. Good coaches are never winging it. They can't make it up as they go. It's written on the paper exactly what is going to happen at what time and for how long.

When I was teaching math, I did the same thing. I made my lessons down to the minute. For example, we might spend nineteen minutes working on a particular type of problem, because that was all the time I had. We had to move on to other things I had to teach that day. So, I had to get as much out of those nineteen minutes as I could. Once you realize how little time you have to get ready for test day, that time becomes precious. You can't waste it. Choosing activities that are most effective in getting results is crucial.

Choosing Activities That Get Results

To teach fundamentals and application in short, high-energy lessons and still give my students time to practice meant I had to choose activities that addressed the priorities I set and would lead to mastery within the timeframes I had. Activities using manipulatives, for example, may be suggested by the book publisher to help my students learn the concept I was teaching, but if they took too much time to set up, I might not be able to fit them into my timeframe. As a football coach, I had to have all my materials out and ready to go before practice. The bags were ready, the helmets were on, and we were ready to go. I had to do the same thing in my classroom. If a lesson using manipulatives helped my students understand the concept, then I had to have it ready before the lesson began so it didn't suck up my time for teaching that concept. I also had to re-evaluate my favorite lessons. As we talked about in earlier chapters, there may be a lesson you like teaching, but if it doesn't teach your students to do the concept or technique that's your goal for the day, it can't be in your playbook.

What worked best for me were short explanations followed by practice problems, breaking down each technique my students had to learn into a sequence that would lead to mastery, and then doing a lot of reps. For example, when we moved on from adding fractions with like denominators and the answer in simplest form, the next skill I would work on was changing one fraction so both fractions had like denominators, but the answer would still be in simplest form. Just a small step. Every new lesson was made with me thinking about making students confident and successful. First, I would work with them on the concept that fractions must have like denominators so they can add them together. I would show examples on the board, then I would show examples of like and unlike denominators and ask, "Can we add these fractions in their present form?" The students would respond, "No," and I'd ask, "Why not?" This is the start of the students understanding the situations they would see in fraction problems on test day. Then I would let them try some problems on their own. I'd walk around and see how everyone was doing. If

The Science Behind the Success

Reduce Cognitive Load, Increase Success

Reducing cognitive load, or the amount of working memory needed to learn new material, can help your students learn complex concepts more effectively. When designing a racehorse lesson, for example, you keep cognitive load to a minimum by limiting the instruction to essential subconcepts then practicing them. If any of the steps need to be done automatically, practicing them until they are mastered reduces the amount of working memory needed to complete that step in the process before applying it to the problem's solution. By sticking to the essential elements of the concept, teaching each step in a process, and providing practice along the way, your students are more likely to be successful (Bruning, Schraw, & Ronning, 2001).

there was a major problem, I would stop the class and help with the difficulty the class might be having. When I was confident that we had the skill mastered, we'd move on. If a strategy for a situation involved a lot of steps, my plan broke them down into the techniques they needed to practice. I had to have this plan ready in case my students had trouble learning them. If they didn't need as much explanation, we wouldn't use every step, but I had to have it ready just in case it was needed.

Getting Started: Getting Your Playbook Ready to Get Results

To get ready to teach situation math in a racehorse classroom, you need to look for high-rep situations. Your goal is to have everyone successfully understand the lesson by the end of the class period. To do that, you need to be prepared to break down each situation your class will face on test day. Your priorities should be based on the most difficult skills they need to learn. Start with the test and ask yourself, "What is the toughest concept? What is the one that is going to trip up my students?" Set that as the priority and give less time to other skills that your students may already know.

To do this, you must have a plan for every day. Almost every teacher has a lesson plan, but in order to accomplish your goal, you'll need a plan for each minute of class to take the best advantage of your time. It also ensures that you are not taking time out of class to decide how something should be taught. No wasted movement on your part means more time for learning. Look at your lesson plan for the day and decide how to break it up so students have enough information to understand what you're teaching but not so much that they lose interest before you stop for practice. Your experience will give you a place to start on this.

Frequent repetition is a time-tested means of mastering fundamental skills. When deciding on the activities for your daily plans, choose activities that will make your students succeed and will give you as many reps as possible. For example, if I knew the book did not have enough problems for the students to reach mastery in a certain skill, I made sure that I had extra worksheets. These extra problems were there if we needed them to reach the level of skill needed to move on. If we didn't need them, that was fine. When I felt we were at the correct level of understanding, we could move on and use our newly learned skill on the next problem. Just because I had the materials didn't mean I had to use them. But I wanted to be ready.

Make sure the activities are moving your class forward and not sideways. Look at your favorite activities critically with this in mind, choosing strategies that ensure your students' time and your time will not be wasted. Use teaching strategies that have a scientific or research base to show they will increase the likelihood that students will understand and master the skills you are teaching too.

The last thing you need to have ready before you walk into your classroom is a way to make sure you and your students know when they have mastered the strategy or technique you taught before you move on with instruction. To effectively monitor and adjust classroom instruction, have quizzes ready to check for understanding after each step of the process you're teaching. Depending on how they do, you can move on with confidence, knowing your students are ready, or provide corrective feedback and reteach, if necessary, if they aren't.

Chapter 8

Presenting Instruction: Making Every Minute Count

"Without properly executed fundamentals, the entire system can break down."
Bill Walsh, Finding the Winning Edge

"Telling isn't teaching."
Tubby Raymond, Former Coach at the University of Delaware

"Success is a project that's always under construction."
Pat Summitt, Head Women's Basketball Coach, University of Tennessee

My goals were set, and my planning was done. Now it was time to set up my class and present instruction to teach situation math in a racehorse classroom. I wanted my students to master each skill they would need to beat the test on test day and gain the confidence that comes from doing something difficult, something no one thought they could do. To do that, I would have to get their attention and their cooperation until they saw their improved grades on that first test. Unlike my football players, they weren't in my class because they chose to be. And many of them had not had success in math in the past. To make all the planning and organization in my classroom pay off, I had to show my students I was committed to their success and give them some relief from their past of long hours of struggling to learn math concepts.

The Science Behind the Success

Setting Up Your Classroom to Increase Academic Engagement Time

How you arrange your classroom and interact with your students has a big impact on academic engaged time. Reducing distractions and setting up quick, efficient routines for nonacademic activities increases instructional time which gives you more time to teach.

Strategies that also have been found to help increase academic engaged time include monitoring students during work periods (checking to see if they are on the right track with the problem they are working on helps this strategy do double duty), going to the student when there is a question rather than having them come up to your the desk, and arranging your classroom so you can see all students wherever you are. Another tip: giving students a few seconds to formulate an answer when you ask a question increases the chances your struggling students will participate in class. (Gettinger & Ball, 2008).

The sense of urgency that I had been feeling since I set my goals and started planning would also be a feature of my instruction and of mentoring my students each day. I don't mean panic. I wanted a high-energy classroom where my students were actively engaged in solving problems and learning skills. To do this I would need a businesslike classroom focused on instruction and practice where success could happen for all my students.

Setting Up my Classroom to Teach the Racehorse Way

Now that I'd set up my lessons to teach in the new method, it was time to set up my classroom to accommodate the racehorse way. The goal in every class was to increase engaged time for learning by increasing reps, or practice. I needed every possible minute of the class period for instruction to make that happen. The obvious place to start was getting rid of distractions. Instead of the office calling my room in the middle of class to tell me a student's lunch money was in the office, for example, I asked my principal to have calls come in at the end of the period or to email me. I promised to get back to the office between classes.

Next, I looked at my routine at the beginning of class. How much time did I spend getting students ready to start? Getting out pencils, books, and etc. can take valuable minutes away from instruction. To be successful teaching middle school math, I had to look at what I needed done during the class period and what I could let go. I found from years past that you can't do it all. For example, I needed everyone's name on the paper but not the period or the date. I had to let that go. I also couldn't waste time while students looked for pencils. I had a can of pencils and handed them one if they forgot. The teachers on my team felt I lost an opportunity to teach responsibility. Learning responsibility is important, of course, but there is more than one way to teach it. My students were going to be taking a lot of responsibility for their own learning very soon, so I could let the pencils go. The good news is that the students who bought into my system rarely forgot their pencils. They learned the routine and were ready to go very quickly in a few days.

Consistent routines also save time. I set up the rules in my classroom so that it became clear to my students right from the beginning of the school year that they needed to be ready for instruction quickly. I also taught them that from the time they entered my classroom until they left, they were expected to think only of math. The students who bought into the system were amazed at how well they were doing when they got feedback on the problems they worked on during their reps and when they saw their grades on the first test. Knowing that they could work on math for forty-five minutes during class, without homework, and still get an A on test day helped a lot in gaining their cooperation. Seeing their results on test day increased their confidence too. Getting results without a tutor or a long night relearning math homework their parents' way meant they were energized when they walked back into my class the next day. They were ready to be successful again. Let me reiterate that last point—they did this all without homework.

What to Say on the First Day of School

One of the skills successful coaches have is the ability to communicate their visions to others, especially their players. This was a skill that easily translated from the field to the classroom. What you say to your students and what you do each day needs to tell them not just what you expect but also that you can help them be successful. What you say and do on the first day of school sets the tone for the rest of the school year, so it's particularly important to get it right. Explaining the racehorse way on the first day of school should get their attention, give them the motivation to buy into your system, and help you all hit the ground running right from that first day.

I knew if I was going to reach my goal of completing the math curriculum by the end of the school year, I couldn't lose a day, so I taught math the first day. I told my students how many days until the first test and explained that we needed to get ready starting that day, developing that sense of urgency I wanted us all to feel. Then I talked to them about what was in it for them:

"When you walk in that door, you will need to be 100 percent focused on math. You are going to learn it now, not later. No one can help you understand the skills you need better than you can learn them here. What would you rather do, work hard in class for forty-five minutes or have to work for three hours out of class with your parents or a tutor? You should enjoy your free time, so spend a focused forty-five minutes working hard now, and you won't need to spend hours doing the same thing over again tonight."

> ***The Science Behind the Success***
>
> **Your Actions Can Increase Engagement**
>
> Students are more engaged when there is a high degree of interaction with their teacher. Your interactive instruction and availability during work times can increase active engagement. Goal-directed lessons taught with frequent opportunities to practice using your feedback and support and provide a sounding board to students who are wrestling with a new problem has a positive impact on the level of engagement and commitment your students make to your class (Gettinger & Ball, 2008).

Right about here, I started to see many of my students' eyes swivel my way. What was I saying? Did I just say they only had to do forty-five minutes of math each day? Come to class, focus, and that's it? Some were still suspicious. Then I told them the rest.

"I am not a big homework person. If you follow the plan, you will not need to spend hours doing homework every day. Learn it now; just think math when you are here. 100 percent math, 100 percent of the time. Don't think about anything else or work on anything else. Then leave the room and give yourself a break from math and come in tomorrow fresh and ready to work hard again. How many people would like to have an A as a final grade? If you follow the plan and work hard in class, we can make it happen. I say 'we' because we will work as a group to improve everyone in math. It's going to take the group to make everyone better. I'm going to show you how."

What was the reaction after I gave the talk on day one? Many students bought in right away. Less work with better results makes most students sit up and take notice. Then I told them the results from last year. Eighty-three percent of the students had an A or B as a final grade. Most of those kids felt they weren't good at math. The rest of the kids had Cs. No one failed. The class got very quiet; I pulled out my playbook, and we got started.

Presenting Instruction with a Sense of Urgency, Not Panic

Once I had their attention, I wanted to give my students that sense of urgency that comes with preparing to learn in the time we had. I did that by creating a crisis. Lou Holtz, the former Notre Dame head football coach, said if there isn't a crisis, he'd invent one. Basically, what he was saying was that every week there was a big problem that the team had to solve if they were going to win on game day. As a coach, I used this strategy and found it to be very helpful for getting my players to focus on a strategy. Maybe the team we were preparing for had a great running game, or they had one player who was outstanding. Taking away their best play or making it difficult for their best player to succeed gave my team a chance to win.

As a math teacher, I used the same strategy. During my planning I had found the situations on the tests that I knew would cause the students the most trouble and made one my crisis of the week. Then I told them, "This is where the test is going to try to beat you. But we won't let that happen because we will be great at this by the time we take the test." I wanted my kids to get the message that learning this strategy might be tough but not impossible, and they weren't in it alone. I was there to make sure they understood the skills they needed to beat that situation on test day.

Urgency doesn't mean panic. Your classroom should run in a calm, businesslike way with everyone's focus on the task at hand. Think of a fire drill. You practice every month. Everyone knows what to do. They may not know whether it's a drill or not—anything can happen—but the routine has been set and is understood by everyone, so when the time comes, your classroom acts quickly and quietly to get to safety. That's the atmosphere you are looking for in your classroom each day.

Getting Ready for Racehorse Practice in My Classroom

Once I had identified the situations, I had to prepare my students for and fit them into the time I had. I wanted to be sure my presentation got as much out of that time as possible. And I had a great model to follow, thanks to Coach Dennis Creehan.

Creehan's racehorse practice is not a passive affair where players stand around listening to the coach expound on blocking and tackling. Some coaches feel if they are talking or demonstrating, they're doing a good job. Creehan doesn't think so. How can someone listen to you and do the task at the same time? Racehorse practices are planned so players can have opportunities to try their moves within the context of the play and perfect the skills needed to be successful. Getting in their reps is key to being prepared. Being prepared, according to Coach Creehan, builds confidence.

Time is set aside, however, for instruction. This is done at a pre-practice meeting with the players. Racehorse philosophy is also used here. No time is wasted, and new concepts are repeated often. The meetings go fast and start on time, and the coach talks with intensity. The coach knows this is where he needs to get his point across to the players. He does this with a fast-paced lecture, repeating himself to ensure understanding. This is the time to teach anything new or different that will be practiced that day.

Then the team heads to the field to practice what the coach just taught (Creehan, 2000).

Coach Creehan likes to teach his drills by mass teaching to keep the racehorse practice going. Players learn more by doing than by watching, so every player does every drill every time. Any corrections for individual players need to be quick and not hold up the whole team. It can be difficult to see everyone at the same time, but the coach will see any player who does the drill incorrectly and will correct their technique as soon as possible, before it becomes habit. Frequent repetitions and early corrective feedback ensures that players learn drills correctly and they are prepared to use them effectively during the next game. Mass teaching also develops leadership in players. Players learn to help each other, as the older players help coach the younger players to improve their skills, acting as positive role models. This helps the team become a cohesive force (Creehan, 2000).

To help the practice maintain this racehorse pace, the coach sets up practice space for maximum engagement. During the team's practice time, there are two offensive huddles. The second group heads to the ball as the first group finishes their play. Players must keep up a fast pace to stay out of the way of the other group. The coach stays in one spot, and the players run past on the way back to the huddle. Creehan knows from experience that coaches sometimes talk too much when correcting one player individually. By staying in one spot, it forces the coach to be quick when giving feedback, making the most important point they need to make without slowing the pace of practice. Practice is only stopped when everyone needs to hear an explanation and correct the same error before their next rep.

The Science Behind the Success

Teaching Motivation

The more academic engagement you can design, the more successful your students will be. So far, we have looked at getting as much time as possible to teach and then setting up lessons that are challenging but with your support, you students can successfully learn. The last factor to consider is student motivation, or willingness to attempt a task. It turns out you can develop strategies that allow your students to increase their motivation too. When you provide challenging tasks along with support and feedback designed to provide information about their strategies, you increase your students' persistence and motivation (Bruning et al., 2011). Success, it appears, is motivating.

In my classroom, my goal for each day was to give quick, high-energy presentations followed by time for practice. I wanted a high-rep lesson for each concept my students needed to practice. Math students learn by doing too, so I needed to get in as many practice problems as possible. Instruction, practice, and corrective feedback were all going to be done in forty-five minutes. Unlike a lecture approach, I wasn't going to spend a lot of time introducing a concept before my students started practicing it. On the practice field in football, each part of a skill must be mastered quickly. We did this by getting in as many reps as possible. If a player was learning a blocking technique, he was going to be better if he got forty reps than if he got ten. To reach my goals, I had to provide as many reps for my students as possible in a class period. That meant I needed to keep instruction time short and to the point with plenty of time for practice problems in between.

To teach racehorse in football, you repeat the most important concept over and over until your players get it. Transferring to the math classroom, I spent my time on the most important things. Explaining them as many times as you can emphasizes key points, so the students know they're important. Explain them to the point where the kids say, "Yeah, yeah, we've got it," and then let them get their reps in.

Football is a game of mistakes, and so is math. The fewer mistakes you make, the more likely you are to be successful. I believe that instruction needs to be focused on the fundamentals, because if they're not solid, students won't be able to apply them to problems efficiently. They can't move forward to application if they give up brainpower on fundamental skills. My goal, then, was getting them so solid on the fundamentals they were confident in handling the situations they would face on the test. They would be thinking about the problem and not the process.

To make that happen, I taught fundamentals within the context of situations. In my planning, I came up with the situations my kids would meet on the test, and I always taught fundamentals to the situations and explained them in terms of those situations. Instead of going over problems and explaining how to do just that problem, which is common when going over homework, for example, I wanted my students to see the bigger picture. To do that, I explained the situation they would see on the test or on problems we did in class, so they would understand why they needed to know a particular strategy and when to use it. They learned to evaluate their errors and explain them within the context of the situations they would face on the test. They began to know when to apply strategies on their own as the reps began to pay off. Confidence that comes from success learning solid fundamentals made applying them to new problems, such as application problems, easier for my students.

Take, for example, teaching mixed numbers. I wanted to give my students a strategy to use when they saw that situation. I might explain it like this: “The hardest problem you’re going to see on the test is a mixed number divided by a mixed number. Learn this strategy and then realize it’s going to take a while to do, so relax and work through it, because you’re going to have to work it down until it’s a fraction times a fraction, then improper fractions times the reciprocal, before you can work the problem. It’s going to come up on the test, so you have to be ready for it.”

This is also a good example of why all that planning is done in advance. When I was setting up my lessons, I had to be sure that I had broken down each technique as far as I needed to so my class could learn each step in the process they would need to beat the situations they would see on test day. Within the concept of dividing fractions, there are many steps that must be understood and mastered. Breaking them down so they could be taught quickly and practiced as much as my students needed to master them also encouraged my kids more. If I could say, “Just learn to do this part well right now,” they were more likely to stay with me and feel better about their math skills when they mastered that situation, one step at a time.

They would need to master those fundamental skills to be able to apply them to application problems, like word problems. Word problems are tough because there are so many situations within each problem, so the strategies I taught were ones they needed to use regardless of the type of problem they were facing. For example, you should never leave a problem without checking to make sure it makes sense. Estimation before you even get started is part of this strategy. My students heard me say this over and over until they were applying it whenever they faced a story problem or needed to figure out an application question during practice. You also need to figure out if you have what you need to answer the problem and know to ignore what you don't need. They heard that often too. Then, I explained, after working the problem, ask yourself if your answer makes sense. Strategies like this helped my students be prepared for the situations they were going to come up against frequently on test days.

Getting the reps in doesn't mean rushing kids when they are learning skills. Once I had given instruction, I had to give the kids time to think it through by working on problems on their own. I wanted to make sure high-rep practice time was quality time. Although it was hard, I didn't talk when they were trying to work. I gave them a chance to think for themselves and work on the technique. People often walked into my room and noticed it was very quiet. It needed to be. When it's time to work, everyone should have time to think through a strategy and learn it without someone interrupting them. Learning to stop talking when your students need quiet to work can be tough but leads to faster mastery.

Monitoring and Adjusting Instruction in the Classroom

Many coaches record their practice to see if their players have mastered the skills they were supposed to learn. Based on what they see when they rewatch the practice, they could decide if they need to reteach and practice the technique again. I used quick quizzes that didn't take the whole class period and didn't count toward the grade. Sometimes I gave a quiz just on parts of the technique. As I explained to my kids, they had to get this one step mastered, or they wouldn't be able to keep going to master the whole technique. If they didn't do well, I knew I would need to adjust instruction and teach and practice it again. But by the time they took the quiz, I expected them to do well. I was disappointed if they didn't. Just like when I was coaching and my players lost on game day, I was disappointed when they didn't do well because I hadn't put my students in a position to win. Unlike football, though, I still had time to get my students ready for test day by reteaching the concept.

Another way I checked to make sure my students understood what I taught them was watching them work their problems in class. While they were doing the reps, I walked around checking problems without interrupting them. It gave me a lot of information in real time, like a film study does for coaches. What you're observing tells you what you need to do next. If they were getting the concept, I could move on. If not, I needed to reteach.

As soon as I felt they were ready, however, I had to move on. I couldn't stay on something they already knew. I didn't have time for that. If they didn't get it, I retaught it. When I had one or two kids who didn't get it, I found time to get more reps in with them. By monitoring this way, I made sure all my kids were ready for the next step in instruction.

Getting Started: Classroom Leadership Makes This Work

In your math classroom, you are the leader. Your vision for your students is the start of your plan for their success. You've set the priorities for what will be taught based on what you know about your curriculum and your students' skills. Then you'll have to make sure to see them through. After I set up my priorities for my classroom, I needed to make sure, on a daily basis, that my "team" stayed on track. I had to make sure that the routines I set up were followed every day by everyone, including me. I set as one of my priorities that all students would do the ordinary things well. Like most of the coaches I have introduced you to, I felt mastery of fundamental skills was the foundation of successful, independent math students. That starts with you—how you set up your class, what you say to your students on the first day of school, and how you back up what you say with what you do.

Organize your classroom and routine to make the most of your time.

Once you've set up your lessons to teach racehorse, it's time to set up your classroom to accommodate the racehorse way. Your goal in every class is to increase engaged time for learning by increasing reps. Using every possible minute of the school day for instruction is the best place to start. Find the time wasters lurking in your day-to-day operations. Look at the typical things that interrupt you during instruction and find ways to minimize them as much as you can.

Next, think about your routines. How much time do you spend getting students ready to start? You may not have thought about how much time your routine takes. This is an area you have complete control over, so set your routines up so you have exactly what you need from your students to get started, and no more. You can't do it all and reach your goals, so decide what you need and what you can let go. Once you set up your routines, stick with them every day. As your students buy into your system, they'll be ready to go very quickly. Consistent routines save time.

Another routine I set up in my classroom was giving all students time to think. That meant when students were working on a problem individually, the room would be quiet, no exceptions. They couldn't talk to each other, ever, and I wasn't going to talk to them. Everyone understood the reason for this rule and that our chances of reaching our goal of beating the test was more likely to happen if everyone followed it. This isn't easy to do. Your first impulse is to remind students of what you just taught them or give encouragement. Learn to stay quiet and watch while your students start using what you just taught them on their own. Finishing a tough problem builds a student's confidence in their ability. Give them the opportunity to experience it.

Set up the expectations for your class based on your common goal.

You've put a lot of time and energy into planning your math lessons to reach your goals. But you can't get there without the full cooperation and engagement of your students. How you present your expectations will determine how much cooperation you will get. Start by explaining the goal to everyone in the class. Everyone must take the tests, so the focus of your class is to get results on test day. Your rules and routines logically follow from that goal. Now it's not about following your rules, it's about everyone working together to reach your shared goal.

Set up the rules in your classroom so it becomes clear to your students right from the beginning of the school year that they need to be ready for instruction quickly. Explain to your students that from the time they enter your classroom until they leave, they should think only of math. Assure them that if they do this, they can learn all they need from you to get results on test day. No tutors or homework required. Your leadership is about making them successful.

Do without homework to increase motivation and save time.

I devote a whole chapter to looking at the research related to homework, including when it works and when it doesn't. But let's face it, kids don't try out to take math. They must do it whether they're good at it or not. "No homework" is the motivator for getting your students to focus in class. If you want your students to buy into your system and get the most out of your planning, tell your students you will teach them all they need to know during the forty-five minutes they are in your class if they give 100 percent of their attention. Most of your students will take that deal.

Present the crisis and the plan for facing it.

You've already identified the toughest problems your students will face on test day. Present instruction to energize your students to beat that crisis. Make sure they know they're not in it alone. You are there to help them be successful. Keep instruction short and teach to mastery. Provide quick, high-energy lessons with frequent repetition until you're sure all your students understand. Teaching this way also helps your students stay focused. The more information they must take in at a time, the more overwhelming it seems. Learning in smaller chunks followed by practice reduces anxiety and gives them the chance to show themselves they can get results on their own. Success builds confidence, which makes facing bigger challenges easier. The result is more learning, more success for your kids, and more confident learners.

Consider your presentation style.

Students need to be engaged as close to the full class period as possible to master fundamentals and apply strategies to solve problems effectively. Do you spend time talking to a student about a problem while everyone is waiting for you to finish? Successful football coaches never stop the whole practice to tell one player one thing. Find another time to talk to them. Corrective feedback needs to be given quickly and privately unless the whole class needs to hear the same explanation.

Monitor progress for student success.

Your goal is for everyone to be successful, so monitor instruction and your students' high-rep practice to make sure they have mastered each step of the process or strategy you're teaching and can apply them to the practice problems. Make sure everyone is successful by the end of the lesson. If your initial instruction doesn't get the results you want, reteach right away. When you're confident your students are ready, give a quick quiz that doesn't count toward their grades to see if you've put them in a position to win on test day. If they don't do well, reteach until they do.

When you've prepared for as many contingencies as you can during planning and used every day and every minute in that day the best you can to get your students to be successful, it can pay off for you in many ways. It worked for me. I finished the curriculum for the year, nobody failed math in my class, and most got As or Bs. More importantly, my relationships with my students changed dramatically. Based on shared goals we worked to reach together, everyone knew they were working to get results on test day, and they weren't in it alone. I watched their confidence grow as they got results. They were proud of their success, not because I made it easy but because they learned difficult concepts with my guidance. And along the way, as you'll see in the next chapter, each class became a team.

Chapter 9

Coaching Your Class: Developing a Winning Team to Get Results for All

"Teamwork is what makes common people capable of uncommon results."
Pat Summitt, Head Women's Basketball Coach, University of Tennessee

"Leadership is a matter of having people look at you and gain confidence by seeing how you react. If you're in control, they're in control."
Tom Landry, NFL Defensive Back and Coach of the Dallas Cowboys

Opportunities for failure are high in any math classroom, and the kids in your class know it. Some learned that lesson early on and are reminded every year. Others have had success in the past but still see every new challenge as a chance to fail. Others know that no matter what they do, they will let their teacher down. They just won't be able to rise to the teacher's expectations. They don't have the background to pass math this year—or any year. Maybe they missed an essential skill along the way. Maybe they just don't get math. They are defeated before they start.

Almost from birth, we all strive for independence, the ability to make our own decisions and do things on our own. Math students are no different. They want to be able to stand on their own two feet and be successful. So, when they can't win, they will do one of two things: give up or fight back. Angry kids will disrupt the class to keep you from noticing they don't get it. Scared kids turn into wallpaper so you won't notice them at all. Defeated kids will wait for their tutors or parents to help them after school and check out during your lesson. Why try to be independent when it's so much easier to let someone else do it for you? These are the games your students are used to playing in school. This year, they will be playing a new game: getting results.

You've set your goal that all students will be winners on the common assessments. Your schedule is set so it uses every minute available to you. You've evaluated your teaching techniques and have lessons ready that will introduce essential skills and provide practice that gets results. You've gone beyond just fundamentals too. Each and every student in your class will have a deeper understanding of math concepts, so they will know the right strategies to use in the right situations. Now what? Your students' cooperation is critical to your success. How do you get your students on board with your plan? You get them to join your team.

Setting up your class to be a team allows you to develop relationships with your students that increase cooperation while they work at a high level of engagement. It helps all your students reach the goals you set for them while they build confidence in themselves, based on the success they can see every test day. To make this happen, you need to communicate a few key ideas. You do this by how you act in the classroom and by what you say to the students.

Teams Are Built Working toward a Common Goal

Not everyone would think of a classroom of students as a sports team. You don't choose to sign up or try out to take math. For me, my class is a team for a very simple reason. We're all working toward the same goal. That's what teams do. Football teams work toward winning the game. Everyone in my math class is there to do well on the test.

> ***The Science Behind the Success***
>
> **Use Challenges to Develop Student Motivation**
>
> Students are more likely to be motivated to succeed when they perceive that the class goals and expectations are challenging but reachable for everyone (Christenson, et al. 2008).

Identifying a shared task is the start of a real team. I started by letting my students know that our goal is to beat the test. Like my players who came to practice to prepare to win on game day, my students and I were going to get ready to get results on test day. Working as a group to defeat a common foe is a unifying experience. My students and I would work together to "win" against the curriculum. The whole idea of us versus the test rather than students versus teacher helped my students and I have the team atmosphere I wanted. We were all focused on that one goal. We would all work together to defeat the test. But, unlike me and my football players, no one was trying to stop us from winning!

Teams Build Relationships That Help Everyone Succeed

You may be asking yourself how you can build relationships with students when you're setting a fast pace to keep up with your lessons. You might even be thinking, "Won't my students resent me?" As a person who has been on many teams, I can tell you the ones that made me feel the best, the most successful, and the most needed were the ones that had the most difficult goals. Similarly, I built the most lasting relationships with my students after I started setting high expectations.

When you teach math like a football coach, you build relationships from the inside out, not the outside in. That is, you don't develop a relationship and then teach. The relationship grows as you reach your shared goals. You won't need to play games or have ice cream parties to try to get to know your students. That kind of artificial relationship fades when kids start to struggle. You create your team through shared experiences, celebrating your wins and pushing through tough situations together, when concepts are hard to understand. Bonds are built when you do your part to help your students be successful.

Everyone has a role to play on a team. Yours is leadership. Leadership in the classroom works best when based on working relationships designed to put your students in a position to win, just like on the football field. To build a successful football team, coaches need to know and understand their players. Not only their skills, but also how they respond to criticism, how they respond to praise, and what drives them to succeed. Bill Walsh made the 49ers a legend in football by not focusing solely on his stars. He developed a plan for each of his players. With his support, his team became legendary. Tony Dungy, best known for being the first black head coach to win the Super Bowl, took his responsibility even further. In his biography, *Quiet Strength: The Principles, Practices, and Priorities of a Winning Life,* he recalls that he never lost sight of the fact that he was "entrusted with the lives of others." He set high expectations and allowed no excuses, but no player on his team needed to worry that he would give up on them while they worked to reach those high expectations (Dungy, 2007).

At football practice, we watched our players and came up with plans to strengthen their weak skills and improve their strengths so they could help the team to the best of their ability. In my math class, I did the same thing by observing my students while they worked. I watched them solve problems. While they were working on the challenges I gave them every day, I learned a lot about what skills they had, how they thought through a problem, what they did when they got frustrated, and what it took to keep them learning and excited about the successes they had. I kept these observations in mind as I monitored and adjusted my instruction so everyone in my class could help the team while improving their skills, confidence, and ability to work. When the weakest member of the team improves, the whole team is better.

Teams Work Best with a Strong Leader

Taking on a common foe doesn't mean you're one of the kids. Kids will never say it, but they are looking for a leader when they walk into your class. Don't be their friend, be in charge. You can be friendly, and you should provide a positive atmosphere, but always remember you are in charge. You are the most important factor in these kids' success. You are the only one in the room whose confidence in your plan and in your group of kids will help them see it's possible for each one of them to learn math on their own this year—no parents, no tutors, just their hard work and determination. One of John Gagliardi's players remembers how much more confident he felt walking on the field on game day knowing that his coach was on the sideline backing him up. Preparation gave him the skills and confidence to play his own game. His coach standing on the sidelines gave him the confidence to apply what he knew (Weeres, 2010).

Most importantly, you must want all your students to be successful. You need to believe in the system you planned to have all kids prepared to get results. The best football teams were

The Science Behind the Success

Setting Up a Motivating Environment

The student's relationship with the teacher goes a long way toward ensuring motivation. Setting up a classroom with clear goals for mastery using challenging activities that focus on understanding and giving students choice in selecting solutions and taking an active role in learning increases their motivation. By providing these activities along with support and assurance, teachers increase student motivation and student success (Christensen, et al., 2008).

not the best because they had superstars. The best teams, like Walsh's 49ers, were the best because the coach set up a system that allowed every player, regardless of their talent, to become the best they could be. John Gagliardi's teams could transcend the fundamentals to a deeper understanding of the game and called their own plays. Dennis Creehan's players learned to help each other during racehorse practice and became a cohesive force on game day. For all these successful teams, it all started with a strong leader who believed in his plan. You need to be the leader that the students will follow. That starts by believing in yourself and your plan.

You must be a strong leader to get things done. You can't just assume your students will just know the material you are teaching that day. You must keep everyone moving toward mastery. You've set up a great schedule and found the most effective teaching strategies available to teach both fundamentals and problem-solving. You need to assure your students that you are not worried about staying on that schedule or getting them ready to be successful. That's under control. Now you need to get them to understand that you can make success happen for each of them.

Be Yourself

Don't be a phony. Kids can see right through that. The coaches I've introduced you to have very different personalities. Bo Schembechler, best known as the head football coach for the University of Michigan from 1969 to 1989, was never considered a nice guy. He was tough on people, but he was a great coach. Bill Walsh was a perfectionist who agonized over every detail of the game. He was tough on himself, but he was a great coach. What came through was their commitment to the success of the team and the growth of every member of their teams, not their faults. Teaching, whether on the football field or in the math classroom, is a personal relationship. You can't fake it, so don't try. Let your commitment come through your way.

Being yourself really hit home when I was a new teacher teaching in an elementary school and I was being pressured to be like everyone else. I had no idea how to be someone else, but I realized I was an expert at being me. I began setting up my system for success using my experiences, humor, and understanding of my students. I learned that when you try to be someone else, your students can tell they are being handled by a phony. In the process, they lose respect for you. Without respect, your authority is compromised and you lose control of your team.

Start with Your Expectations

As the leader of this class, you set up your classroom so you and your students are all working toward a common goal by having your expectations clear in your mind and in your students' minds. You want to show them that everyone can succeed if you all work together. Start by just laying it out on the first day the way we talked about in Chapter 8. Explain that you're all working to do well on the test. To do that, you're all going to come in prepared, you're all going to work hard, and you're all going to be successful. That's what you expect to happen. As their teacher and their leader, I explained to my students that I wanted the same thing they did. They wanted to be successful, and I wanted them to be successful. I told them, "I'll come in every day, and I'll have the lessons ready. You come in and do everything I have planned every day, and you have a chance to be successful."

While you make this goal clear to your students, you are also making your goals front and center for you. Every day you are going to be working toward putting your students in a position to win. You're not wasting time with anything else. All talk in the classroom is about the work. Those discussions you have about not having materials? They stop happening. Think about it. When you're chewing a student out for not being prepared, what are the other kids doing? Nothing. No learning is happening. You're not working toward your goal. And your students will notice.

Lead by Example

If you say you are running a winning classroom but you're complaining about the same things they hear in their other classes, like disciplining a student for five minutes because they don't have supplies, you lose your kids. They see right through that. You waste time, something you told them they can't do. But if you start teaching from the time they barely sit down, now you are showing them how your class is different. You're reaching your goals for the day, and they can reach theirs. *"This is what we have to get done today,"* you explain, *"so we can* ***defeat the test."*** I like to give a quick pep talk when I start. I might throw in at the beginning of the lesson, *"This is a pretty tough concept to learn, but I know we can do it."* I let them know I have confidence in them, and they get excited and become engaged in the lesson right from the start.

You must come into your class every day with high energy and be ready to go. You are showing your students how a successful, well-prepared, businesslike classroom is run. If you consistently run your class like this, the kids come to understand that this is the way it's going to be. They might as well hop on, because the train's leaving the station. There's no panic, just good time management, something each of your students need to learn to do well in school and, eventually, when they get their first job. The kids are learning how to work successfully by your example.

The Science Behind the Success

Fairness Increases Motivation

Christenson, et. al. report when students are successful with challenging tasks in a classroom they perceive as fair (where the focus is on mastery of material instead of rewarding high grades) and receive support from their teacher to be successful, their motivation increases Christenson, et al., 2008).

Set the Rules, but Not Too Many

Fairness is important to middle school kids. Researchers who study successful middle school classrooms have found that students have less engaged time in classrooms they see as unfair.

It makes sense when you think about it. Why try hard if only a few can succeed? Why not disrupt the class if your teacher isn't committed to your success? Since your classroom is set up so anyone can get an A if they master each step of every process you teach, you just increased their academic engagement and their trust in you. They have a fighting chance this year, and you are going to show them the way.

The rules you set up in your classroom also show your students you are committed to fairness. If you are going to treat everyone fairly and they are going to treat each other fairly, not a lot of rules are needed, so don't go overboard and take away their pride. The rules in your class should support the fight against your common foe, the test, so don't spend too much time on classroom control. Set a few classroom rules that are needed to reach your shared goals, and that's all. They should reflect respect for the learning of others and ensure maximum engagement while in the classroom. John Gagliardi expected his players to follow the Golden Rule and support their fellow players, with the older and more experienced guys mentoring the new players. Tony Dungy learned from his father, a teacher, that the choice to do well was ultimately up to each individual. His father never took away that decision from his students. By preparing them for each test, he increased their confidence and their chances of success, but it was up to them to make that choice (Dungy, 2007).

I used to have more rules than I have now. They were about how students should act while in my classroom, such as using appropriate language, keeping your hands to yourself, and following directions. Now I have two rules. First, come into the classroom ready to work. Give 100 percent of your focus on math for the whole class period. Second, treat everyone the way you want to be treated. That's it. I keep them very busy during the class period, so there isn't much downtime to get in trouble. Granted, it can be difficult for some students to focus and get started. Usually just walking near a student who isn't working is enough to get them on track.

This gets easier as the year goes on and your students experience success each day. Once the students see that they can do the exercises you give them and learn each step of a math process, they gain confidence. They develop a positive attitude based on their own hard work and great test grades, which will lead to more hard work and more success on the next test. This circle of success will feed on itself to higher levels of achievement, motivating students to try ever-more difficult tasks. Why act up in a class where you're successful?

Science Behind the Success

Motivation is Based on Beliefs

A student's beliefs about their ability has a direct impact on their motivation. Students, like all of us, are always deciding whether they are good at something by how they perform and the feedback they get back from others, like their teachers. If they see themselves as successful, researchers have found, they are more likely to try again and to stick with a difficult problem in the future (Bruning, et.al. 2011).

Your Students Have a Role to Play in Becoming a Team Too

Helping each other makes teams stronger, but how do the kids on your team help each other? The main way they help each other is by staying focused and reaching their goal. You're expecting your students to work on each problem and do each rep that you have planned for that day. Your rules let your students know what you expect from them. Following them consistently helps everyone get results on test day. If you're a student sitting in class and you don't care, you might start talking to someone else. Now there are two kids not learning. Even if you don't go that far and check out during class, you won't learn what you're being taught, and you slow down the class's progress because you are going to need more repetition. That takes time away from everyone reaching their goal. If you're disruptive, you take your teacher's time away from the class. Now no one is learning.

As some of your students start contributing to your discussions and asking questions, it helps others see what it looks like to learn and get results. If they're focused and answer the questions and you're positive in how you respond to them, even when they're struggling with the task, your classroom becomes a safe place to ask questions and try more difficult problems. Kids who aren't confident in their math skills see how you respond to their classmates and start answering questions too. Soon they feel more positive and get excited because they can join discussions. That helps the group feel good about their success. Instead of fading into the background, even the most reluctant students realize they help each other the most by working hard, doing well on the test, and learning the skill so everyone can move on. Now you don't have to stop instruction to help the ones that don't understand or backtrack to get everyone to mastery as often. Their success helps everyone.

Being told that you are part of a team will probably not be enough of an incentive at first to gain the students' cooperation. That first test is a few days or weeks away. Some incentive to help them give it a try is not getting homework. As a student, you are less likely to be disruptive if there is something in it for you. You are more invested if you can say to yourself, "Here is my chance to do well with the least amount of work." If you have found math difficult in the past, you will want to avoid it as much as possible. You are also more likely to stay focused if you can see the end. If you know you can be successful by staying focused for forty-five minutes and then you get a break, mentally you can probably handle that. You can handle working hard for a short amount of time. What you can't handle is doing math all day and all night.

This is one more way that you help students decide to buy into your system and make the choice to be part of your team.

Fostering a Positive Attitude

Cooperation is simple for your students. Participate and treat others fairly. What you do with that will determine if your plan succeeds. It's one thing to say everyone will be treated fairly. Your students are going to be watching your actions to see if you mean it, especially those that haven't decided to buy in yet. To gain the cooperation of your students, foster a positive atmosphere in your classroom. I don't criticize a student for a wrong answer. Math is all about trying strategies to reach a conclusion. There are a lot of ways to get there, and we've already talked about how you must evaluate those answers to see what make sense. That problem-solving won't happen if you shut a student down during this process. As my discussions grew from our shared goal, to get results, I showed my students by my actions that as long as they were working on a solution, it was worth considering. That made it safer for my students to take the chance of being wrong and evaluate their own mistakes, which is a big part of learning.

To foster a positive attitude in your class, you have to be positive with your students. In addition to sending the message that everyone is going to be successful today, you need to show that everyone's questions will be answered or addressed without criticism. If the answer doesn't come out right, you all work together to figure out what happened. There's no judgment on the quality of the solution. In addition, when running a high-energy class, you need to be supportive of students as they learn what they need to reach today's goals. There's a relaxed but businesslike atmosphere. If something's funny, you laugh about it. If you and your team do well, you celebrate (for a few seconds). Your excitement about their success helps them be excited too.

Keep Your Poise

While everything you set up will reduce the need to confront you in class, there will still be students who will try to push your buttons. This is a time when you need to keep your poise. No matter how you may feel about what a student said or did, you can't get emotional. You must keep your composure. There are two very good reasons why it's crucial for you to do this.

First, you want to leave the lines of communication open with the student who confronted you. If you develop an antagonistic relationship with them, you lose their trust, and your working relationship goes out the window. It makes it much more difficult to work with them after that. Your lessons will likely fall on deaf ears. Responding negatively to a student also impacts how other students react to them. If you belittle a student or give them a hard time, it isolates them from the class. Now others will be less likely to include them in discussions or respond when they have an idea. That student won't have much reason to cooperate. They won't be motivated to treat others like they would like to be treated. The team you were trying to build will likely have one less member.

The second reason you need to keep your poise is your class will be watching. If you lose your temper or even respond sarcastically, the other kids see that and realize they can get to you. You could lose the respect of the class. Even if you must remove a student from the room, you must stay businesslike and calm. Keeping your poise shows your students that you are still in charge of the class and will keep them working toward the goals you all need to reach to be successful on test day.

How do you keep your poise when a student pushes your buttons? Do this by sticking to the facts of the lesson. Clarifying the question by saying something like, "What are you asking me?" redirects attention back to the task at hand. Another way is to ask the student how their comments or question is related to the problem the class is working on. Again, you are redirecting the attention back to the lesson.

Classroom leadership, I find, is a lot like coaching on game day. Even though coaches can be criticized these days for not showing their emotions, someone must be ready for the next play. Someone must be thinking about the next thing coming down the road. While it may be very satisfying to roll your eyes or raise your voice when dealing with tough situations with kids, you need to consider what the consequences will be for you and your class later on. Treating students in a businesslike way makes it clear to the student involved that you're looking to solve the problem. It tells your class that you are still in charge and focused on your mutual goals. It shows them they can trust you.

Getting Started: Teaching How to Work

I got a thank-you letter from a student a few years after I taught him in math. The one line in the letter that stood out was "You were the first teacher that taught me how to work." Here was a student who realized that hard work was a good thing and who felt the pride that came from being a part of a team that worked hard together.

Grouping is never easy. Any teacher must evaluate their situation in the moment and decide what's going to work. I had heterogeneous grouping, so I had kids who were advanced and others who were weak in math and study skills. Because of this, I had to have enough reps so they were all solid on the skills. Then I had to teach the kids with weak skills how to work. Once they learned to do that, they made a lot of progress. Aside from the skills themselves, learning how to work well was the most important thing for them to master to become successful learners. After they learned it, they could look around the room at the successful learners and realize they knew how to be successful too. I recognize these strategies may not work for every student or situation. Some students have their own reasons for not wanting to be successful. They may be battling their parents and trying to make a point by not doing well in school, for example. But developing a team in my classroom worked for me, and I believe it can reduce the amount of time it takes to manage your class and make a lot of learning happen in a day for you too.

How did I teach him how to work?

- *By my actions*— I showed him what it looks like to come to class every day and be ready to be successful.
- *By showing him the steps in logic*— Math is logical. There are strategies for figuring out if an answer makes sense. I gave him permission to try, even if the answer might not be right, and apply those strategies on his own.
- *By allowing him to decide to cooperate*— It was his choice, not my pushing, to become part of the team. I wanted him to learn that by making the decision to participate, he could have the same chance to be successful as anyone else in my class.
- *By giving him the incentive to try his best*— By not giving homework, he learned to go all out in class, rest up, put in the same effort the next day, and be successful again.

Another student once said to me, "You make people happy," after I gave the class a challenging problem. I knew that if my students persevered, they would be able to figure it out. They didn't know it, though. I wouldn't let anyone quit on themselves. I encouraged them to try applying what they were taught that day and see how it would come out. I was ready to remind them to use strategies to see if their answers made sense, but I held back from telling them what they were. I wanted them to experience the pure joy of figuring something out on their own. The students were so excited about successfully finishing the task by themselves that I could tell this situation didn't happen often enough. What does this mean? I think students are tougher, want to work harder, and will push themselves more than we give them credit for. We should not take away learning experiences because we think they can't handle them. We must believe in their success.

Rudyard Kipling wrote "For the strength of the pack is the wolf, and the strength of the wolf is the pack." People working as a group can reach higher levels of achievement than any individual can. You and your students will begin to feel this in your classroom. As everyone improves, the entire class improves. Lessons start running more smoothly. Less time is wasted as everyone learns the routine and gets going right away. Soon everyone in the classroom is achieving at a higher level. Students get along because they feel part of something special. And you gain confidence because you were the reason for their success.

Chapter 10

The Case for "No Homework"

"All progress has resulted from people who took unpopular positions."
Adlai Stevenson, American Statesman

"Never venture, never win."
Sun Tzu, The Art of War

A tough challenge of this plan is the "no homework" component. You really have to step outside your comfort zone to not give homework. Can students learn math skills without doing homework every day?

Not only can they learn more math in less time the "no homework" way, but this approach will also help your students feel more in control of their lives while they're *not* doing homework. They will build pride and the confidence that comes from getting results from their own efforts. And did I mention that you'll come to school every day feeling more rested and as excited to see your students as they are to see you?

Of course, it doesn't have to be uncomfortable for you. Why should it be? Your playbook was set long ago. You spent time making sure you have as many instructional days as possible squeezed out of this school year. Everyone on your team understands what you're trying to do and maybe even helped you find the time you need to get results for your students. You set a goal that all your students are going to get great grades this year. Okay, so maybe no one thinks you can reach it. Your pride won't let you back down now. Not after you spent the summer planning like never before, leaving nothing to chance. You now know how precious each instructional day is, so you've analyzed the skills and concepts needed for your students to get results on each test day this year and set priorities based on their needs. You planned each lesson so fundamentals are racehorse taught and the strategies your students need to get math at a deeper level are practiced under your supervision, so your students can walk into your classroom on test day and get results for themselves. In other words, you are now ready to teach your students to master basic skills and develop a deeper understanding of the math concepts they will need on test day to have the confidence to plan their strategy and get results on their own.

As a math teacher, I can tell you how important homework is to my colleagues. I've heard all the arguments for giving it. The most important on being students need it to make sure they have enough practice to learn math skills. Homework is something that is so sacred in education that we very seldom even consider not giving it. But there is strong evidence that it's not adding much to the amount students learn. In this chapter, we'll explore the way I look at homework, using what I learned from the successful coaches I read about. Do they load it on to get their players ready to play on Saturday? At what point does the extra work give diminishing returns? Then I'll give you the arguments for going "no homework" and provide you step-by-step directions for setting it up in your classroom. I'll also look at the research in education related to homework, especially related to math, so you are ready to show how "no homework" can work with your students. So, get ready to go "no homework" and watch this year's math students become the best you've ever had.

If Kids Could Learn More Math in Less Time by Not Giving Math Homework, Would You Stop Giving It?

Bill Walsh's 1987 San Francisco 49ers started the season on a bad note. The team lost their opening game to Pittsburgh. However, they rebounded during the season, despite a players' strike, finishing the season thirteen and two. The 49ers ended their regular season with a win over the LA Rams, 48-0. But it came at a cost. In his book, *Building a Champion: On Football and the Making of the 49ers,* Walsh recalls that because the Rams were a big rival team in their division, he kept his starters in too long. Instead of getting the game over with quickly when they had it in the bag, he kept throwing the ball to prove a point to himself that his team was better than the Rams. Everyone on the team was drained physically and emotionally and did not recover well. Then the week before the playoff game, they practiced too long and too hard, trying to perfect their deep-passing game. The players were beat physically. The first game in the playoffs was with the Minnesota Vikings. On game day, the Vikings took it to them, and the 49ers were flat. The team, one of the best of the decade, just could not play up to their usual standard. Bill Walsh knew they'd spent too much energy on the Rams and over-trained for this first playoff game. Because of this, his players were not able to get results.

Tired, Confused, Angry Students Don't Learn Much

Let me tell you why I stopped giving homework. I knew that students could not study math all day without a break. I saw this happen myself for years. Students refused to do their homework or were too tired to give their best effort in class and didn't experience success. What typically happens when students don't do well in math? First, their parents might get them a tutor to help them with their homework. When their grades don't improve, we keep them in the room at lunch and recess to do more math. Now these students will spend up to two hours a day on math, around 8 percent of their day. These students know full well that someone will explain the lesson again that night when they get home, so why pay attention to the teacher in class? I saw too many kids with that glazed look in their eyes, waiting for class to end. I felt the kids have the life sucked out of them with nothing to look forward to but day after day of math drudgery. And in the end, I knew they didn't understand math any better. And so did they. The next day when they came to class, they had that same defeated look in their eyes, and now they were exhausted and discouraged from studying all night. Like Walsh's 49ers before the 1987 playoffs, they were flat. How were they going to get results on test day?

The answer to the problem of underachievement many times is more of the same: more time on math, more homework. This is the safe way out. Just keep piling on the homework, and it looks like you are doing your job. But the kids will come to resent you because they have no control of their lives. They will lash out or check out. Not only do they check out of your class, but they also check out at homework time too. After all, if your teacher, your parents, and your tutors make all the decisions, your pride will say, "I need more control." Since them learning math is more important to the adults than to these students, a great way to get control is not to do your homework. I am sure many students are angry with the well-meaning people in their lives because of this loss of control.

No Homework: Why Your Students Need It

But I think a greater problem than the loss of control is the loss of pride. One of the major ways people motivated me as a child was to challenge my pride. I never wanted someone to think I couldn't do something. A simple "you can do better" was enough to get me going. But as a teacher, I often heard my students say, "I can't do any better," or "I'm bad at math," or "I don't care." All that extra work taught them was they couldn't learn math on their own. On top of that, their parents or tutors didn't always know the way I showed them in class, so their homework was done a different way than the way I taught them. Now they were confused and didn't learn either strategy to mastery. Even worse, perhaps an over-involved parent did all their homework for them. The message that comes through loud and clear to these kids is, "I must not be smart enough to do this on my own, because someone is doing everything for me." Whoa, I had *a lot* of work to do. Not only did I need to teach these kids how to work, but I also needed to teach them to believe in themselves. That is why I feel it is so important to set up a classroom where students will develop pride in themselves because they can learn math on their own and see the results for themselves.

How Your Class Is like the Panama Canal: Keeping the Mud from Sliding Back

This would be a great time to ask, "Why do I need to teach kids how to work? Why do I need to teach them pride?" It reminds me of the story of the Panama Canal. When the French first tried to build the canal in 1881, all they did was dig. But the more they dug, the more they had to dig. The mud just came sliding back. The men were dying of malaria too. They were sick, and they were tired. They were going nowhere—they weren't getting results. Then, the United States took over in 1904. The first thing they did was take care of the mosquitoes. They drained water, sprayed the ponds, put up screens, and gave the men quinine. These efforts kept their workers healthy and better able to work every day. Next, they solved the engineering problems with effective strategies to keep the mud back. Those problems needed to be solved before the work could begin. Healthy workers using the latest engineering advances got results. Alert, confident kids who are taught to be successful get results too.

Just like the builders of the Panama Canal, you can't just keep digging until everyone drops from exhaustion and gives up or goes home. You must figure out what the problems are and solve them before you can hand out the shovels. Doing math homework caused three major problems for my students. First, it added to my students' confusion about math. Did you ever have to learn a phone number that was only different from another one you already knew by a number or two? How many times did you ask yourself, "Is that the right one"? Things we must learn that are similar but not the same are harder to master and feel confident about it. So, when my students' parents taught them a different way to do the problems I assigned, it was new learning. Similar, but not the same. We talked about Creehan's racehorse approach to learning to skill mastery. If I taught my students one way to solve a problem, for example, and their parents taught them another, they didn't get more practice. They got something else to learn. It didn't look like the same thing to my students. They just learned a lot more stuff in one day, something I worked and planned to keep from happening. The result was confusion and discouragement, not mastery or pride.

We need to develop our students' pride in themselves so that when they are faced with the battle between one method or another, they can choose the right path. This brings us to the second problem math homework causes. Our first thought when faced with a student who is struggling in math is to do *more*—more hours studying, more problems to solve, more practice with skills. But football coaches know that players cannot practice all day. As Bill Walsh found out, long practices make players flat and do not help them play up to their potential on game day. And as Tony Dungy learned, the quality of our preparation is more important than the amount of time we have to get ready (Dungy, 2007). Taking up all their free time with football does not lead to high performance for players. Practices are planned to get the most out of the time you have. Coaches also know that they can't teach their players when they go home. It must be done on the field. They must watch to see if their players are learning the skills correctly and have practiced enough to master them. Coaches have a sense of urgency to get the job done right then. They don't send work home for parents to do. They don't suggest getting a tutor to teach blocking skills.

Nothing is more important than the time spent in class. A football team works together with complete focus during practice. Every minute is accounted for. There is no wasted time. The coach's main goal is to teach the skills and knowledge needed for game day so everyone can perform them flawlessly to get results. If a drill takes a large amount of time and does not provide a large gain in learning, it's eliminated. Players leave the field satisfied with something left in the tank. All players have met some success and are mentally getting ready for the game. Because they gave total concentration to football for two hours, they now have time to give to the other parts of their lives. Mentally, it is something the players can handle. The cognitive load is not too great, and they can see for themselves they are ready. If a coach said, "We need to practice for five hours tonight, then you have to go home and study game film for two more hours," for example, the practice would be terrible. The players would go into low gear, nothing would be gained, nobody would look at film, and I'm sure the team would lose that game. It just doesn't work. Why do we think it will work in the classroom? Let's let the kids come up for air.

"No homework" develops that same sense of urgency in your students that football players feel at practice. They will learn quickly that their success depends on learning the skills now. Today is the day. You will review, but the bulk of the learning takes place now. Also, their pride will develop because they know they are the reason for their success. They learned it on their own. Nobody had to do it for them. Almost every student can mentally handle forty-five minutes of all-out math, especially when they know that's it for the day. They will walk out of your class with pride in their math ability based on the success they had that day.

One exception to "no homework" is studying for quizzes and tests. But as students develop pride in their work and confidence in their ability to learn, they will have the motivation to prepare for quizzes and tests independently. And it won't take them long. Because your quizzes will be short and usually test one skill, they will not need to study for hours at home. Because you developed this sense of success in them, they come into the quiz feeling prepared, so test anxiety fades. Their grades go up. Great grades on tests show your students they can be successful, which leads to better classroom work as new concepts and processes build on what your students have learned. Better classroom work leads to higher achievement and better grades. Higher achievement leads to more success. Success builds pride. Confident students with solid skills get results.

What Do You Say to a Discouraged Student? It's All in the Delivery

As we discussed in Chapter 8, when you introduce this to your students you need to get their attention and keep it. Ask your students what their first choice is for a reward. Mine usually said "free time." We all value free time. It belongs to us to do with as we wish. It also gives us a chance to rest, regroup, and pull together all the ideas that have been floating around in our heads all day. Now you sell them on the idea that if they work hard now, they are done for the day. There will be an end in sight. Remember, they walked in feeling tired, confused, and angry. All their past experiences up to then have taught them they're not good at math and hours of drudgery are coming when they get home. I don't believe that you can get the level of attention and engagement needed to get results if you give homework.

"No homework" pulls this whole plan together. Since I also am working from a sense of urgency, I need to make sure I get my students' attention right away. When I talk to the kids on the first day of school, I ask, "Would you rather work on math for three hours and get bad grades, or would you like to work hard for forty-five minutes and get an A?" Active engagement, coming right up. I just drained the canal and killed a few mosquitoes. Now I can hand out the shovels. My students see that I am thinking of their needs and not just loading up the work. In return, I tell them, I will give my time and attention to each of them, so everyone can get results on test day. Now your students see you as a true leader, someone who can help them be successful. Someone they can trust to follow. It is no longer them against you. You can now develop a climate of winning and make the test, not you, the opponent. You and your students are on the same team, and, as we saw in Chapter 9, that's the start of your students becoming a team.

No Homework: Why You Need It

Up to this point, if you have been following the steps in each chapter, you have put in a lot of time preparing for that first day of school. You've developed a sense of urgency in yourself by setting a goal that probably looks impossible. You've looked at the number of days you have to teach and then found some more. You've prioritized your lessons so the most important fundamental skills can be learned to mastery and you still have time to teach your students to start applying those skills with confidence. You've set up your classroom to increase academic engaged time. Your head's in the game. You're ready. "But," you are probably thinking, "my students are not football players. They don't want to be here." Good point.

Math Students Didn't Sign Up for This

Unlike football players, your students didn't sign up for your class. They are in your class because they have to be there. Many of your students don't come in with high expectations for getting results themselves. They aren't eager to get started. While kids who make the football team have a lot of motivation to show up for practice to win on game day, your students are in your class because they were assigned it. There isn't a lot of excitement in their eyes that first day. As we've seen, many of them have had bad experiences and don't consider themselves good math students. Others hate school in general and, by middle school, are good at disrupting class, so no one finds out they don't have a good grasp of math fundamentals.

You are about to find out how to turn that around. While football coaches may have to temper their players' enthusiasm at the beginning of the season, you must build it. But around the part of the season when football players start to get discouraged as they lose some games and coaches must work up some enthusiasm in them again, your classroom will be hitting its stride. Your students will have learned more math in less time than ever before while building pride in their ability to learn on their own. By the end of the "season," you will have shown each of your students that they are math stars—by not giving homework.

We would probably agree that motivated students are easier to teach. What better motivator is there than free time? That gets your students' attention. Now there is something in it for them. I'm sure some kids are thinking they are getting the best of this deal. But because your instructional goals and the strategies you want them to learn stay under your watchful eye, they will learn more math, one skill at a time. Because they will come into your class less fatigued, they are more alert and ready to learn, too. And because your leadership has shown them your commitment to their success, they will grow to trust you. How many students have you had who are watching to see if you mean what you say before they will cooperate in class? As you build your team, you get the chance to show them through your actions that you are committed to their success. You've addressed their need for quality instruction through your planning, and they got results. You've reduced some of the fatigue they feel at the end of the day through not giving homework. You've shown them they can get results if they trust you. They're much more likely to be motivated to learn in your class.

Teaching Math Without a Safety Net

It's scary to look yourself in the mirror and say, "I am the reason for the success of my students." The safe and easy way out is to give gobs of homework, get the kids a tutor, or tell parents to work more with their kid at home. Most parents of struggling math students have been there before. They know the drill. No one would blame you. You did all you could. But nothing will have changed for your students.

If you are like most teachers, you are very confident in your knowledge of your subject matter. You know math. If you have been following the steps in each chapter, you know planning and classroom leadership. You're ready to go. Now you just need to have the confidence in your ability to help students get results on test day for your team—every day, in forty-five minutes a day. Like a good teacher, Bill Walsh set up his training program, leaving nothing to chance. He wanted those skills taught to mastery because, in the end, it was his responsibility to get results on game day (Walsh, 1998). All coaches know they are accountable for the results their team gets on game day. It's a fact of their lives and why they are such good examples for me. That's what gives them that sense of urgency I've been talking about.

"No homework" will develop that same sense of urgency in you that you've instilled in your students and that successful coaches feel, because you are ultimately responsible for the success of your students. All your students must learn the skills you planned today. They must learn fundamentals to mastery today. You must have a positive attitude about their ability to learn what you're teaching today. You must check how all your students are understanding and help them go from understanding to application today. You must make sure that all your students know you are committed to their success on test day every day.

Lastly, because you won't stay at school all night correcting homework, "no homework" will give you a mental break. This is the most important part of the day. Just as your students can't do math all day and all night, neither can you. You need time for your family and time to rest. And you need to be fresh for class the next day. Trust me: you will need to be at your best to pull this off. There will be no downtime in a class period, and you need to have the same level of clarity in every period you teach. But you can also relax, knowing your planning was done before the school year started and the lessons are focused to get results on test day.

Does "No Homework" Work: Or What Do You Say to a Nervous Parent and a Skeptical Administrator?

What do you say to parents, other teachers, and administrators who say, "What? No homework?" Let's start with parents.

Most parents will back off when you tell them their kids will study for quizzes and tests. The rest will be convinced when their kids start getting results. Once their kids show them 100 percent on tests—the goal you set for your students when you started this process—parents will most likely be on your side. Also, they aren't paying for a tutor anymore, so you're saving them money. And what parents don't need some time off from arguing with their child about doing their homework?

Of course, some parents simply must have homework for their kids. It's an ingrained habit by the time they get to you, but your goal is to get your students a rest break. Tell their parents that their child's homework is to play outside for an hour. Explain to them that physically fit kids are more alert, so getting a break will help their kids pay attention better in your class the next day. You're not just preaching. You need them alert to get the benefit of your instruction. This is important to help their child get results in your classroom.

Apparently, I'm not the only one that has noticed that tired kids don't learn much. Researchers into the mental health of students have found sleep deprivation is a major contributor to poor learning. The National Sleep Foundation recommends that children between six and twelve years of age get between nine and eleven hours of sleep and adolescents should get between eight and ten hours of sleep per night, and many students aren't getting it, even in elementary school. It's estimated that as many as 31 percent of elementary-age students don't get adequate sleep, and it's even worse for young adolescents, with as many as 71 percent getting less than eight hours of sleep (Laracy, et al., 2015). More importantly, for us as classroom teachers, lack of sleep is a major contributor to poor school performance. Both decreased sleep quantity and quality impair academic performance. Sleep duration alone took up 64 percent of the variance, or influence, on "cognitive processes that impact academic functioning, including working memory, short-term memory, and attention," according to psychologists Laracy, Ridgard, and DuPaul in their review of the sleep-deprivation literature. And Kidwell et al. cite research that getting less than the recommended amount of sleep by even thirty to sixty minutes can result in noticeable reduction of a child's functioning in school. It appears students need sleep more than they need homework.

Parents can be encouraged to have students use the time they would be doing math homework to get more rest.

Then again, nothing beats authentic application. Let parents know that you suggested to your students that they help balance the family checkbook, measure a recipe, or figure out how much flooring is needed to cover the kitchen. Suggest that they begin to include their children in these everyday experiences when they occur. It won't happen overnight, but every day their kids will move toward being more confident in their ability to do math on their own. Their parents will see their kids become the independent learners they need to be to achieve success throughout their lives. In other words, the arguments will cease when you and your students get results.

Does Homework Help Students Get Results? What Does the Research Say?

Explaining this to your principal or your colleagues may be a bit tougher. Part of *our* work ethic is giving homework. If you're like me, you started giving it when you did your student teaching. You know the arguments for giving it too. It's a necessity to practice basic skills to help master them. Doing homework improves a student's work ethic, study skills, planning, and organization while teaching personal responsibility. Homework is a tradition in education. But to make an objective decision about whether to give homework as part of your lesson plans, we need to look at what the research says about its effectiveness. Let's start with the first question we should always ask when we want to win on test day: Does giving homework get results?

The amount of homework has been increasing in the United States in the past thirty years, especially since the beginning of the 2000s with the introduction of No Child Left Behind. As demands for higher achievement levels have increased, so has the amount of homework, with many elementary students receiving hours of homework in some schools, even though it has not been found to improve achievement for elementary students. Some researchers suggest that the loss of downtime, and increased amounts of sleep deprivation and stress we put on students hurt student achievement more than they help. The amount of homework interferes with a student's family life and time for other activities, such as sports, extracurricular activities, or recreation and family time. Alfie Kohn, noted lecturer and social scientist who has written about education and parenting, has suggested that the current state of affairs burns out even high achievers (Kohn, 2007).

There is some disagreement about whether homework is effective, however. A review of the literature by Harris Cooper, a social psychologist, researcher, and professor at Duke University who has been studying homework since 1987, (Cooper 2006) found that when you look at short-term results such as unit tests or class grades, some benefit could be found for homework for secondary students. The studies done since 1987 in his review show that homework had a small positive impact on unit tests and standardized achievement tests for high school students and a smaller effect for middle school students. But many studies conflict with this conclusion. Even some of the studies reviewed by Dr. Cooper found that giving more homework had an adverse impact on achievement. Also, it had no positive effect on elementary students through grade six and even caused those students' attitudes toward school to deteriorate. The effect on achievement, when it exists, then, is only in the short term. A lasting impact on achievement has not been found, Cooper concludes. (2006) He's not alone in that conclusion.

In his book *The Homework Myth: Why Our Kids Get Too Much of a Bad Thing,* Alfie Kohn reminds us that a positive correlation with homework doesn't mean homework caused that difference in student performance. He reviewed the data Cooper and his colleagues compiled and found that there were so many methodological flaws in the included studies that conclusions should be made with caution. Positive correlations, when present, may not be because of the effect of homework but the result of a third factor that was not considered in the data, such as student motivation or socioeconomic factors, for example. The data on homework's value is inconclusive at best, he reports (Kohn, 2007).

Another factor that has an impact on test scores and achievement is the quality and effectiveness of instruction. The homework controversy is international. Many policymakers look to the results of the Trends in International Mathematics and Science Study when examining the impact of homework on math achievement. When students in Western countries like the United States and Germany started falling behind students in Asian countries, the fact that they were getting less homework than their Asian counterparts was assumed to be the reason for the slip. The cry went up to increase homework to improve our students' achievement. Ulrich Trautwein and Olaf Kollar, of Germany, noting that strong empirical support for increased homework to improve achievement had not been found, decided to look at seventh graders and see if homework assignments improved math achievement (2002). They noted that many things can impact the effectiveness of homework on achievement, such as prior knowledge, intelligence, socioeconomic level, the motivation of the student, and the school they went to, so they controlled for those things. They then found a positive correlation between frequency of homework if the assignments were short and math achievement tests. Lengthy assignments were associated with lowered achievement. *It also made no difference on achievement if teachers checked homework or not.* Like Kohn, they point to the need for more control of these other variables, along with separating out the different features of homework, such as length, type of assignment, and frequency. In a later study, they reiterate their conclusion that long homework assignments do not improve achievement and suggest that the effectiveness of teaching methods should be considered in future research (2003).

Is there a point of diminishing returns for homework? It appears that the answer is yes, and we seem to have hit it. Despite the steady increase in the amount of homework that students have been given in the last thirty years, overall achievement has not improved, and some studies show that U.S. students perform less well than students in other countries who get less homework. So how much is too much? For students in elementary school (kindergarten to grade six), no amount of homework is beneficial, and it has been found to harm their attitudes toward school in general. High school students are the only group that both sides of the homework controversy agree benefit from homework, and even they see a point of diminishing returns. Cooper reports in his book *The Battle Over Homework: Common Ground for Administrators, Teachers, and Parents* that doing more than six hours of homework *per week* across all subjects is not beneficial for high school students. In a study of international math achievement, however, increasing the amount or frequency of homework as a way to improve achievement *actually hurt overall returns,* Kohn reports (2007). Like football players who have been trained too hard, it appears our students become flat when we load up the homework.

What About Quality Instruction?

One of the things about homework research that is seldom talked about is that even when positive, the effect of homework on achievement is relatively small. Quality of instruction is rarely studied, but homework researchers sometimes refer to it as a reason for the impact of homework. Trautwein and Kollar, for example, noted that how teachers give homework is often related to the instructional approach they were using, confounding the results of the homework studies. Was it the homework they were assigned or the instruction their students received that led to student improvement? It seems likely that the "other factor" that Dr. Kohn suggests may be causing the correlation, when it occurs, between homework and achievement is quality of instruction. Homework researchers who did rule out the quality of instruction found, he reports, that homework had no "meaningful effect on achievement at all." "Put differently," he states, "the research offers no reason to believe that students in high-quality classrooms whose teachers give little or no homework would be at a disadvantage as regards any meaningful learning" (Kohn, 2007). One high school interviewed by Kohn even stopped giving homework to their students when it was found that it had no impact on the results of their high-stakes assessments.

Kohn interviewed teachers as part of his research, and many did admit they didn't find homework that helpful to their students' achievement. Some of the teachers quoted in Kohn's book found that when they stopped giving homework, their students were more engaged during class and began to show more interest in exploring a topic after it was covered in class discussions, even in high school. One teacher admitted that when his instruction improved with experience, he didn't need to give homework anymore. Others preferred to provide assignments that could be done in class, such as short reading assignments that could then be discussed or written about with teacher supervision.

What about In-Class Supervision?

As we've seen, research is inconclusive for the effects of homework on achievement. But I wanted to know how it stood up to the practice I gave in class. The limited studies that compare homework to in-class supervision find homework has a small positive effect on achievement in high school. There is less effect in junior high and none in elementary school. The positive effect in the upper grades may be due in part to the variability in teacher behavior during in-school study, often providing passive supervision while students finished assignments as the intervention (Cooper, 2006). However, when giving homework is compared to providing skills instruction supervised by a teacher, a lot of that effect goes away. In fact, up to grade six, in-class supervision has a stronger correlation to achievement than homework. There is a point of diminishing for homework even in high school, after which achievement and attitudes toward school deteriorate. Researchers looking at the same data admit that even when effects are positive, they are small and disappear when student characteristics and effectiveness of instruction are ruled out. Trautwein and Koller (2003), among others, suggest future research should include instructional practices to help us understand the role of homework on student achievement. If my experiences are any indication, high-quality instruction using every minute of the instructional day beats homework for getting results on test day.

So, what do you tell your colleaguor your principal about the research about homework? There is no evidence that giving homework improves achievement, as opposed to grades or other short-term measures. It's not better than good instruction, and if you are in an elementary school, it's no help at all. It can even damage your students' attitudes toward school in general. In middle school and high school, it amounts to a small part of the improvement but only if the instruction you are giving in class is top notch. And how are you going to find time to do that if you are going over homework in class?

To Assign or Not to Assign? You Decide.

As we said, "no homework" is the toughest part of the program to accept. It's so ingrained that homework is necessary for kids to learn math that to consider not giving it feels very uncomfortable. Because it flies in the face of traditional wisdom, it can also cause friction with your colleagues. But a strong relationship between time spent on homework and achievement has not been found. Research is inconclusive for the effects of homework on achievement. The amount of time spent on it doesn't make that much difference on tests and skills development. Effective instruction is likely to provide more academic gains than homework does. Your students are tired, frustrated, and unmotivated. You're tired of teaching tired, frustrated, unmotivated kids. So, what do you do?

To help you decide, let's go back to academic engaged time during the instructional day. How much time are you spending managing homework? How many minutes does it take to collect it? How long does it take to deal with the students who don't hand it in? How long to record the grades? Are you going over it in class? How long does that take? Is it helping you to teach situations going over specific problems? When you're explaining it to the students who don't understand, what are the other students doing who did get it? How many meetings are you having or emails are you sending after school to parents whose kids didn't get their homework done? Would that time be better spent on instruction? Time spent managing homework is often a time of low engagement. Imagine what you could accomplish in that time giving direct instruction or in-class supervision.

Next, ask yourself why you are giving homework. Commonly, it is given to provide practice and repetition. It's a double-edged sword, though. If students don't understand what you taught, they may ask their parents or their tutor, who might explain it a different way than you did. Now it's not practice or repetition. Now it's a new thing for them to learn and understand. This is one of the main reasons I went “no homework.” I didn’t have time to reteach my strategies the next day. The next problem to consider is what the students who don’t need the repetition are going to do. They may ignore it. Now you must spend time following up with them and their parents to enforce your homework requirement. This eliminates your prep time and harms your relationships with your students.

Are you giving homework to see if the students understand the concepts you taught? This may be helpful if the students complete it and don’t get help from their parents, but you have no way of knowing what happened last night when that homework was done. As we’ve seen, sometimes the homework is so guided by others that you can’t even be sure the student did it. Well-meaning parents can end up doing a lot of homework because it’s not worth the fight at home. Even if they think they’re just helping, they can help too much and guide your student to the point where they don’t need to think the problem through for themselves. They’re just following along. In a survey completed by Harris Cooper, approximately half of the parents surveyed admitted to helping their child get their homework done faster in ways prohibited by the teacher, including doing the homework for them. Using strategies to test knowledge during instruction and giving time to work in class with your guidance can give you a read on whether students understand the concepts you taught better than homework can.

Some teachers give homework to cover material they didn’t cover in class. By now you can see that losing control over the way your students learn material is a problem. Did someone teach them their way? Did their parents do it for them or guide them so not much learning occurred? When so much of math depends on strong foundational skills, you don’t want to leave mastery to chance. Besides, if you are going to be held accountable for the results your students get on tests or state assessments, why not make sure they get the best instruction available: yours. This way you are in control of your students' learning, and you can have pride in your students' success.

If You Decide to Go "No Homework": First Steps

1. *Plan this first.* Going "no homework" isn't something you do midyear. Plan for it right along with your lesson plans before the school year even starts. If you haven't made your instruction as effective and efficient as possible, you are not ready to tackle the homework decision. When it is clear that you are using every minute of the instructional day effectively, the homework decision will be easier to make.

2. *Understand where it fits into this system of instruction.* Academic engagement is not just about planning your lessons for maximum effect. It is also about your students' motivation and their commitment to learning. Not giving homework is a powerful motivator for your students, especially those who have not had positive experiences in the past. These are the students who are facing a very long day working on a subject they are convinced they will never be good at. By reducing the commitment you expect from your students to one class period a day, you not only have their attention, but you also gave them a reason to cooperate. Now there is something in it for them.

3. *Let the students be your guide.* One of the things to consider when deciding to go "no homework" is to observe your students. What is the class attitude? Are you seeing tired, unmotivated kids? Are you spending a lot of time keeping them with you? In many middle schools and high schools, students help their families by caring for younger siblings or working after school. By providing them a way to learn what they need to without adding to the drudgery of their day, you can increase students' engagement

while getting better results. It also levels the playing field for those kids with more responsibility at home.

4. *Plan your introduction of "no homework" for the first day.* Introduce this as an option to students on the first day of school. Much of the school day is out of the control of your students, so you are sure to get their attention when you discuss "no homework" as an option. Explain it as a choice. By making a commitment to one class period per day, they can get the best grades they will ever have in math, get their parents off their backs, and get some breathing space. Make sure they understand that in return you will give them your full attention and make a commitment to each of them so that they will be successful.

5. *Be ready to explain to colleagues and parents why you are not assigning homework.* Colleagues should hear the research about homework. In middle school, it has little effect on student performance; effective instruction has a lot more. Instead of grading homework and discussing it in class, you want to use that time for instruction. It's a time commitment you want to use for intense instruction to get all your lessons in and taught to mastery. Parents need assurance that you are committed to their child's progress. They need to understand that you will be teaching their children the math they need to be successful on test day, that you are committed to their child's results. After the first quizzes and major tests come home with the best math grade their child has ever had, parents will be on your side. Watching their child gain confidence in their ability to learn math and apply it when it's needed is something most parents will appreciate.

Game Day

Every Lesson is Your Game Day

Chapter 11

Coaching Yourself: Getting Your Head in the Game

"Leaving the game plan is a sign of panic, and panic is not in our game plan."
Chuck Noll, Head Coach, Pittsburgh Steelers

"Indecision is fatal. It is better to make a wrong decision than to build up a habit of indecision. If you're wallowing in indecision, you certainly can't act—and action is the basis of success."
Marie Beynon Ray, Author/Editor

Have you ever been totally panicked and didn't know what to do? Did you freeze or get angry? Our bodies prepare us to run or fight by pumping a double shot of adrenaline into our bloodstreams when we're stressed or scared. The only problem is, either way, our brain shuts down. We are running on instinct unless we have planned and practiced for the worst. Freeze or get angry; this is what your students are going to see unless you prepare mentally each day.

As we've stepped through each part of the process I use to teach like a football coach, what it takes to be successful has become more and more personal. Setting goals that no one thinks you can reach is all about keeping your expectations high for your students. Asking yourself what it will take to make that happen and having a playbook with plans for each lesson helps keep the focus on your students' success too. You're ready to keep expectations high. Not giving homework gives your students a vested interest in following your lead each day. Now it's all in the presentation. And that's about you.

Classroom leadership, in other words, is all about you. Your goal is to keep yourself from panicking and have the calm, businesslike attitude it takes to guide your students to success on test day. Your attitude will come through no matter what, so make sure it's in line with your goals. This is the last piece of the puzzle that will get you ready to teach so almost every student in your class is successful. What you focus on and the attitude you bring to the classroom determines your success each day. If your attitude determines your success, how do you get your head in the game?

Every Lesson Is Your Game Day

Running a high-energy classroom takes an enthusiastic, high-energy teacher. Every lesson is your game day, so you must be prepared to reach your goals for your students with a sense of urgency but not panic, knowing that you can make it happen. With the right mindset, you can walk into your classroom every day to make success happen for your students.

It's easy to get overwhelmed and let panic set in if you only focus on problems. In the book *Football Coaching Strategies,* Glen Mason, head coach at the University of Minnesota between 1997 and 2006, addresses how to deal with tough times. Mason struggled with a team that had had losing seasons for six years. He could easily have let panic stop him in his tracks when he saw all the problems waiting for him, so how did he get past the worry of being in over his head and get started, eventually making it to the Big Ten Championship four times? He focused on the things he could control and looked for progress every day (Mason,1995).

What Do They Need?

Mason started out doing a thorough assessment of the needs of his new team. He saw what was working and what needed to be fixed. He didn't dwell on those problems, though. They were just the starting point. Once he'd identified the problems, it was time to move on to solutions. Next, he set his goals, keeping expectations high. As he explains, you will never get to the top if that is not your goal. Like all successful football coaches, he sees the full responsibility for the success of the team squarely on the coach's shoulders. If the coach can't see the team winning, neither will the players. To do this, he feels that the coach needs to be ready to control what they can and not spend time on what they can't. If you don't have control of the players' abilities, the team's facilities, the amount of money, and etc., there is no sense wasting time worrying about it. Instead, he advises coaches to spend time on preparing the players. Their training is something a coach can always control (Mason, 1995).

Coaching Yourself

The coach's leadership determines whether the team gets results. Because the responsibility for the success of the team is on the coach, it becomes very important, Mason feels, that coaches don't lose faith in themselves. It's easy to focus on what will happen if you fail, but that doesn't help you reach your goals. There will be tough days. Solutions take time to work. But the coach must not waver from their plan to turn the team around. They must believe in themselves.

Mason feels it is important that the coach reminds the team what their goal is and makes them believe they can reach it every day. The coach's attitude has a lot to do with that, he feels. Enthusiasm is important. Being positive shows the team you believe in them. This is not always easy to do. When a team is struggling, there are plenty of people to remind a coach what the problems are. University alumni, administration, and the press all like to analyze and second-guess. If this enters a coach's mindset, the team will not improve. Being negative, according to Mason, continues the trend.

One of the best ways to turn around a losing team, he feels, is to emphasize improvement. Mason believed coaches should work on making everyone better every day. He asked players each day, "How can we be the best team we can be?" If you spend your time on improving, he feels, the games will work out. High expectations are more reachable when everyone sees small improvements every day. It's up to the coach to point those out.

We've already covered a lot of the things that teachers have control over. You have control over your planning. If you've been following along in this book, your playbook for each day is ready to go. You have control over how you instruct your students. You know you can reach your kids by staying focused on your goals and doing what it takes to reach them. It will help you to pull hard in the right direction by keeping you focused on what is important. It will also keep you from looking at certain students as kids who can't succeed and look at them the way coaches look at athletes. Like all our coaches, Glen Mason looked at the skills his players had to determine what training was needed to help the team win. Improving his players' skills was within his control, so that became his focus. This made sense to me, so I tried it out in my math classroom. After a thorough needs assessment, I looked at each student and asked myself, "How can this student get 100 percent on the test?" It was a tall order at times. It can be overwhelming if you focus on all the roadblocks in the way of that goal. I had to find a way to keep panic and worry from getting in my way. I decided that if I wanted to have a positive attitude that would come through during lessons, I needed find a way to make reaching goals fun.

Making Success a Game

Think of a student who isn't doing well in your class. Now you ask yourself what it would take for that student to get an A on the next test. Like you, I've had students in my class who had very poor math skills. I decided to make getting them ready for test day a game. I asked myself, "What can I do in my time with these kids to give them the skills to win on the next test?" As a test for myself, I looked at a student with many academic problems. If you'd asked the members of my team who had the same student, that goal could look difficult or downright impossible. There was a long list of skills this student didn't have, both in math and the other classes. If I focused on what the student couldn't do, it could look downright impossible for the student to get an A. Instead, I asked myself what skills they needed to master to get an A on the next test. We worked on those skills, and that was all. I wasn't going to concern myself with the next test. This student got a 92 percent on the next test, missing an A by one point. My colleagues were surprised, and I felt pretty good about my ability to get results for my students. I must admit, winning that game was fun, both for me and my student.

Football coaches aren't always impressed with the players who come out for the team. They may have more interest than skill. Not only that, as many of our coaches found out, but also lack of success can lead to bad attitudes, so even talented players don't give their best. Football coaches can fall into the trap of complaining about their players, but they're the only players they've got, so complaining won't help. The best coaches get started figuring out how to get each player ready for game day. Putting their players in a position to win becomes their goal. Deciding what they need to teach to get their team ready to be successful helps coaches keep a positive attitude. It's the same in class. They're the only kids you have, so you may as well figure out how to make them successful. You can't do anything about a student's home life, but they're in your class for forty-five minutes every day, and you have control over what happens there. How you teach and what skills you decide are crucial will determine each student's success. Even if you fall somewhat short of the goal, it's a lot better than a student barely grasping the material and not being ready for the next lesson. If a student is in there working hard and getting results, it helps everybody. You make that happen.

Focusing on short-term goals keeps panic away.

Don't drive yourself crazy thinking about the whole year of tests. If I had thought about all the things my students didn't know and what I would have to teach that year, it would have been a lot more difficult to stay positive. I decided to stay focused on one test at a time. Like a football coach, I thought "first downs." That one test was a first down. If my students and I could keep getting first downs, we would be in good shape. My short-term goal each day was to work on skills to make my students successful. What do I need to accomplish today? No more than that. Staying focused on my game plan helped my class stay focused and successful. Staying focused on working just today's lesson, *I* could stay enthusiastic and positive.

It's easier to be positive when you see success. Before that may be a leap of faith.

For my football team, staying positive throughout the season can be difficult. While they may have started out the season with high expectations, losses may start to erode their confidence. It was the opposite with my math students. Many of them were coming in with pretty low expectations on the first day of class. Each day I had to show them that they could learn each step of each process they needed to be successful on test day. Once I got them through the first test and they could see their results, my math class got more positive, and that continued throughout the year. They could see for themselves they were getting results. They started to believe they could be successful math students.

Your students aren't the only ones who need to see that progress. So do you. One of the best ways to turn around a losing team is to emphasize improvement. Work on making everyone better every day. By spending your time working on improving your class by mastering each skill each day, the tests work out. Celebrating those successes each day helps you and your students stay positive. You've reached another goal, and that feels good. High expectations are more reachable when everyone sees small improvements every day, including you. By focusing on improvement, you, like your students, gain confidence in your ability to be successful.

Before that happens, it may take a leap of faith to believe in yourself. It's an attitude. If you believe it, it can happen. If you don't, it will show in your teaching, and you won't reach your goals. If you truly care about everyone being successful, the students will buy into your plan. Since they will be taking their cues from watching you, they need to see your belief in yourself to make success happen and your belief in them to get results. You can say all the right things, but if you don't really believe it, that will come through.

You must believe that you are the best person to help your students be successful. Tell yourself, "Everyone is getting 100 percent because of me." As long as you keep it to yourself and let it just come through in how you teach and relate to your students, it will help. Saying something like that out loud sounds like you don't believe it, or you're just bragging. This is just your attitude about how you will teach. Your pride is on the line here. That's what it was for me. I was tired of being told we couldn't get through all the curriculum and have everyone get As. It was going to happen. Then I found out how hard that would be. That could have been the end right there. But it was something that meant a lot to me, so I kept going. I told myself I was going to be successful. There were kids who thought they were terrible at math, and I said, "Well, that was then. This is now. It's going to be different this year." And it was.

You must be willing and able to take that accountability, however. That's the scary part. Once you step over that line and say, "It's all up to me," there's no other place to put the blame, and you're on your way to having the right attitude to make your students successful. Taking responsibility for what goes right and what goes wrong in your class allows you to decide how to fix problems so everyone can be successful. As you monitor and adjust your teaching, you see improvement. You see that you made success happen for more of your students than you thought you could. Seeing those first test scores shows you your success too.

Taking care of yourself needs to be in your game plan.

Coaches know they need to be rested and mentally sharp for practice. Practice is where each game is won or lost. The long hours they put in, even at the high school level, watching film and planning the best plays for game day are important. It gives them lots of information to prepare their strategy for game day. But they are not putting in the hours to be a martyr or to show up someone else. No one cares that a coach spends eight hours breaking down film. All that matters is the information they get from the work. The hours put in are for results because those are the only things that count. There comes a point, though, just like homework for students, when preparation leads to diminishing returns. Successful coaches know they need to be at their best when working with players on the field. Working late hours may keep coaches from being their best during the most important time: teaching players during practice.

Like with coaches, working long hours can interfere with your ability to stay positive and high energy when teaching. Think about it. If you spend hours designing your lessons and preparing for the next day and you're the last one out of the building at night, how are you going to feel in the morning? When you walk into your class the next day, will you be tired? Irritable? How long before you start to feel resentful? After all, you put so much time into making your lessons the best they can be. Now your students aren't even paying attention. It wouldn't be surprising if you start to feel angry with them for not appreciating your efforts. But none of that will help you be the kind of leader your students need you to be.

You need to be sharp, rested, and ready to go at math time, your game day, every day. Teaching in a high-energy classroom is very intense. You need to be mentally and physically ready each day. Staying up late working just wears you out if you do it on a regular basis. You need your students to be sharp and enthusiastic to be at their best, and you must be mentally and emotionally ready to communicate your enthusiasm to them too. Don't work late at school doing tasks that don't help students. If you follow my planning system, you will have everything ready, so long nights won't be necessary during the school year. You won't be grading homework, either, so you gain a few hours there too.

Give yourself permission to take a mental break each day to do something that you enjoy. Work out or take a walk, connect with friends, cheer for your kids at soccer practice, read a book for fun. Eat a balanced meal, get plenty of rest. Taking care of yourself improves your chances of success on your game day tomorrow.

Getting Started: Preparing for Tomorrow's Game Day

Teaching and coaching are demanding and challenging jobs. But because they're so hard, you can get excited about them when your students do well. Knowing you made that happen, seeing your hard work pay off, can boost your confidence and your love of teaching. That's what I found. The bonus is that your students will see you are honestly excited about their success too. They'll know how much you care when they get results. After all, doing something challenging and being successful is a great feeling. Your students will feel that when they get their tests back. So will you.

Believing in your students' success is intertwined with your belief in yourself. You took on the accountability. That was your leap of faith in yourself. You came up with the plan to make them successful. You took full control of your students' learning. Each small step they make each day shows how your plan works. It shows your leadership. The excitement you feel when your students make a connection during a lesson or get results on test day is your pride in action. Your attitude is communicated each day to your students as you all get results together. That's all about you.

Chapter 12

A Winning Season—And Your Best Year of Teaching Ever

"Doubt yourself and you doubt everything you see. Judge yourself and you see judges everywhere. But if you listen to the sound of your own voice, you can rise above doubt and judgment. And you can see forever."
Nancy Lopez, Professional Golfer

"Action may not always bring happiness; but there is no happiness without action."
Benjamin Disraeli, Prime Minister of England, 1874–1880

I started this book by suggesting that football was the best-taught subject in school. I wanted to prove to myself that the principles I used in coaching were also important to teaching. It is more likely that successful coaches are, in fact, good teachers. Successful coaches use many of the principles of good teaching. In fact, at least one coach we talked about, Dennis Creehan, has degrees in education. The overlap between the two made me realize that while good teaching practices work to make successful football teams, the experiences that coaches have can help educators face the challenges we face now and those we will likely face in the future. Regardless of where you work, whether in a public school or private, parents expect we will teach their child to mastery. They want their child to be a successful, confident learner. That is most important to them, and that is what they will remember, long after their child leaves your classroom.

As a coach, I learned how to make my players confident and successful because there is one advantage coaches have when teaching their players: they work on a short timeframe. It doesn't take long to find out if what you are doing is working when you coach football or any sport. The proof is there at the end of the week. The length of time you have to prepare your team is short. The time you have to address problems is short too. There is no time to admire the problem. There's also no time to be paralyzed by fear. My experiences making the plan, teaching the plan, and being successful taught me to trust what I see and trust what my data told me. That sense of urgency I had as a coach taught me to act based on what I see. As a teacher, I learned to plan and teach effectively. As a coach, I learned how to get my head in the game, assess, and act, taking full responsibility for putting my players in a position to win.

Building a System to Blank Out the Noise

I wanted a way to gain control of the results in my classroom, where it was up to me. I was a teacher and a football coach. I liked doing both those things, and I wanted to be the best I could be at both those things. Using coaching to guide my teaching gave me a complete philosophy, a very big picture to guide every aspect of teaching. It gave me a focus for my planning, a mindset to prepare to teach each day, and a way to reach my kids and show them they could be good at math *on their own*. When I got tired or began to get discouraged, when the pressures from those outside my classroom tried to intrude on my game plan, I remembered the successful coaches I had read about and stuck to my vision, reviewed my goals, and renewed my commitment to help my students be successful. I knew I was right, and my students' grades each test day showed both them and me they could get results. I didn't panic and stuck to my plan.

Not that sticking to the plan was easy. It wasn't easy to be calm and enthusiastic in front of my kids each day. I got tired some days. I wondered if I had done enough when I planned out the year. Did my playbook break down tough skills enough for my kids to understand now, today? Would I know when it was time to move on to the next concept when they were with me? I stayed alert to the way my students responded to the lessons and the answers they got when they worked problems in front of me. I presented the lessons each day with as much repetition and problem-solving as I could pack in, so my students mastered each skill and were ready to pick up where they left off the next day without doing homework. And as we sweated together through tough concepts and celebrated our victories, my class became a team.

It wasn't easy to look in the mirror and say, "I'm responsible for the results of my students." I had to take that responsibility, and by the time I had finished my plans, taking each big idea and making sure it was broken down to its component skills, I knew I was the best person to teach them math. No tutor or parent was going to do it better. They couldn't get in my head and know how to work my plan. Sure, I was saving my students' parents some expense by firing the tutor. More importantly, I was giving them the opportunity to have quality time with their children instead of struggling with them to get their homework done.

I was also giving my students a real gift: confidence. Our self-esteem is valuable because it is so hard won. We gain it by doing very difficult things, facing challenges, and reaching goals we never thought we could reach. I gave my students that opportunity. They had the chance to challenge themselves to be good math students. They could see the results for themselves each test day. They could see it in class when they tried to solve problems using the strategies they'd learned. I was there every day guiding their journey and reassuring them that even tough problems could be solved. I didn't reach everyone, but for many of my students, it was the first time they ever got to see themselves as independent, successful learners. It made them happy in the last place they ever expected to be happy: in math class. Together, we changed their perceptions of themselves from being overworked, tired, unsuccessful, and angry about that to students who could trust themselves to get results even when learning was difficult.

You Are the Expert at Being You: No One Else Can Do It Better

We all bring our unique experiences, talents, and personalities to teaching. No one else teaches quite like each of us. We are who we are, and that's always going to come through. My experiences as a coach and the reading I did to become a better one gave me a unique perspective when I realized coaching and teaching have so much in common. A lot of the pressures are the same, but even more important to me was how the coaching perspective answered so many concerns we teachers have. Planning and classroom discipline top the list for many teachers. And the answers were staring me in the face every football season. If you've followed the chapters in this book, you should have a clear vision, specific goals, and a detailed plan to meet them. You have a different perspective on classroom management and leadership. I made some controversial decisions about what I needed to teach, how I spent my time to make my goals, and what I could let go. Letting go of homework lightened the load considerably and helped my students improve their attention in my class. It was the best motivator I could have asked for. Based on my experiences in my classroom, I am convinced that if I had left out any piece of this plan, it wouldn't have worked.

Putting Yourself in a Position to Win

Now it's your turn to decide what you will do. While the decisions for planning and running the class may seem obvious, put together, they led me to my goal: results for all my students. While you may want to do just part of this plan, they are all pieces of the same puzzle and were all important to reaching my goal. To put yourself in a position to win, you will need each step, especially when it comes to homework.

Developing goals based on your vision is a leap of faith in yourself.

Developing your vision and setting student-centered goals to reach it may seem like a tall order when there are so many obstacles in your way. Doing this not only helps you prioritize your planning, but it also means you are going to focus on the things you can control and ignore that things you don't have any control over. Your goals keep you focused on the reason you got into teaching in the first place: helping students succeed.

When deciding on the activities for your daily plans, choose activities that will make your students successful and give them as many reps as possible.

Instruction needs to be of high quality to be effective. If you've taught math for a few years, you probably know what techniques work best and need to organize your plans a little differently. If you're new to teaching math, using teaching strategies that have a scientific or research base will increase the likelihood that students will understand and master the skills you are teaching and will help make sure you haven't wasted your time and your students' time. From all that information out there, choose the strategies that you can do efficiently in your classroom and help you reach your goals of providing as many reps needed to learn skills and opportunities to apply them in a class period. More importantly, if you have strategies that are near and dear to you but don't help your students make progress, they have no place in your playbook.

Understanding by Design's planning backward so my students could be successful made sense to me. You may be very happy with how you set priorities for lessons. Maybe you liked a few things I did, but on the whole, you feel solid there. For me, planning was essential to getting results, and prioritizing lessons was a big part of that. I had to know what I was going to do each day. I couldn't wing it with my favorite lessons or ease back a day. I saved time by breaking skills down in my summer planning instead of figuring it out that night and going back the next day to try again. I needed my students to get to mastery in the shortest time possible, so I had to be prepared for whatever my students would need to handle all the situations they would face on test day.

You don't have to give up your personality to teach math like a football coach. In fact, you can't. What you can do is know your lessons so well from your planning the previous summer that you get what's covered done effectively, but the way you introduce it may be different, depending on the group. When your planning is tight and you're confident in your knowledge of your plan, you have the freedom to present in a very personal, high-energy way: yours.

Behavior management using the Golden Rule works in class too.

Teams develop a cohesiveness when the members unite to fight a common foe or to reach a common goal. Individuals want to be part of a team where their needs are respected, and they can contribute to the team. Behavior management can take a lot of time and is usually based on compliance or cooperation with you. Making the class about getting results on test day instead of compliance took me out of the equation. We were all there to make sure everyone was successful, including me. We teach our students by our actions as much as by our words. By being confident enough to allow my students to decide to be cooperative, I increased my chances they would become members of my team. Being a mentor to your students may improve your relationships with your kids, give them the motivation to become a strong, confident member of your class, and help everyone get the most out of your instruction. Using the Golden Rule as your class management helps develop a calm, businesslike environment where the focus is on learning math.

Meeting your students' needs by not giving homework improves motivation.

Speaking of motivation, what increases a student's interest in your class? Certainly, success does. It's tough to leave class every day knowing that you just don't get it or only understood part of what you were supposed to learn. If your homework is designed to see if you understand the concepts taught that day, but you don't get to it until hours later, how much will you remember? Monitoring and adjusting instruction when your students are right in front of you and checking for understanding right away helps you ensure all your students have mastered concepts and can apply them successfully to the situations they will face on test day and in the future. So why assign homework? After all, if you're going to ask your students to make a big commitment to staying focused in class, you need to give something back to them, like some of their free time. You don't want them to come to class flat tomorrow either. It's better for you to have refreshed, motivated kids. More learning happens in less time that way.

As teachers, what gets us up in the morning and rewards us at the end of the day are the responses from our students. Imagine a classroom where your students are with you each day. They come in and get ready on their own. They sit up and smile. You get a lot accomplished in a real, authentic way that builds your team. You won't need special activities to build self-esteem and motivation. You'll watch it happen as their math skills develop and they beat tough situations. As the school year progresses and they gain confidence, they'll join in discussions more often or try a new problem on their own. There will be some challenging situations for you too. But you'll keep your poise when, for example, you're faced with kids with poor skills or a chip on their shoulder. It'll be easier when you make the solutions a game. By focusing on improvement instead of problems, you'll keep your head in the game. And no matter how tired you are at the end of the day, you're happy to go back the next day, knowing no one will be able to help your students get results better than you. You can have this be your daily classroom experience—I did. If you follow these examples, set your goals, and are prepared, you can get the results in your classroom so everyone wins!

Acknowledgments

When you set a goal that no one thinks you can reach, it helps to be surrounded by supportive people who believe in you. First, we would like to thank our families for standing by us. Thanks to Marge, Mitchell and Davis Evans for your love of mathematics and your support during this project. Thanks to Robert and James Naunas for your encouragement and every-ready tech support.

Many thanks to Brenda Ringer for patiently reading our beta version, helping us find an editor, and providing vast amounts of encouragement throughout the writing of this book. Thanks for cheering us on.

Grateful thanks to Stacey Kucharik at Polished Print, the first person to say, "publishing date." We thank you for bringing our book to life. And to Alex Ellis, thanks for going above and beyond proofreading our book, suggesting improvements in wording and even doing some fact-checking for us. We are grateful.

Lastly, I'd like to thank the students I coached and taught over the years. Your success is my reward.

For more information and resources, join us at our website: evans-naunas-results.com.

Get inspiration and tips to teach math like a football coach, using strategies for planning, leadership, and building relationships with your students without giving homework.

You will also find worksheets to calculate instructional days and start your planning.

Thank you for reading this book. If you enjoyed it, please consider giving us a positive review.

References

Bruning, R. H., Ronning, R. R., & Schraw, G. J. (2011). *Cognitive Psychology and Instruction* (Fifth). Pearson.

Christensen, S. L., Reschly, A. L., Appleton, J. J., Berman-Young, S., Spanjers, D. M., & Varro, P. (2008). *Best Practices in Fostering Student Engagement in Practices in School Psychology* (Fifth). NASP.

Collison, J. (2001). *No-How Coaching: Strategies for Winning in Sports and Business from the Coach Who Says "No!"*. Capital Books.

Cooper, H., Robinson, J., & Patall, E. (2006). *Does Homework Improve Academic Achievement? A Synthesis of Research, 1987-2003*. Review of Educational Research.

Cooper, H. (2007). *The Battle Over Homework: Common Ground for Administrators, Teachers, and Parents* (Third). Corwin.

Covey, Stephen R. (1989). *The Seven Habits of Highly Effective People: Restoring the Character Ethic.* Simon and Schuster.

Creehan, D. (2000). The Wing-T from A to Z: Installing the System, Volume 2. *Coaches Choice Books*.

Dungy, T., & Whitaker, N. (2007). *Quiet Strength: The Principles, Practices and Priorities of a Winning Life*. Tyndale.

Gettinger, M., & Ball, C. (2008). *Best Practices in Increasing Academic Engaged Time in Best Practices in School Psychology* (Fifth). NASP.

Kidwell, K. M., Hankey, M., Flores, D. M., Van Dyk, T. R., Lundahl, A., & Nelson, T. D. (2015). Sleep and School, Strategy Brief. Lincoln, NE: Student Engagement Project, University of Nebraska-Lincoln. *Communique*, *43*(5), 12. Retrieved from http://k12engagement.unl.edu/sleep-and-school.com

Kohn, A. (2007). *The Homework Myth: Why Our Kids Get Too Much of a Bad Thing*. Cambridge, MA: Da Capo Lifelong.

Laracy, S., Ridgard, T. J., & DuPaul, G. J. (2015). *Sleep and School Functioning: Guidelines for Assessment and Intervention* (Vol. 43.8). Communique.

Legendary Gagliardi Decides to Call it Quits. (2012, November 20). *Star Tribune*. Retrieved from http://www.startribune.com/sports/179961001.html?refer=y

Mason, G. (1995). Football Coaching Strategies. *Human Kinetics.*

Retirement Rumors Again Swirling, Gagliardi Faces First Losing Season since '67. (2012, October 11). *Star Tribune.* Retrieved from http://www.startribune.com/sports/gophers/173823501.html

Tomlinson, C. A., & McTighe, J. (2006). *Integrating differentiated instruction and understanding by design: Connecting content and kids.* Alexandria, VA: Association for Supervision and Curriculum Development.

Trautwein, U., & Koller, O. (2003). The Relationship Between Homework and Achievement: Still Much of a Mystery. *Educational Psychology Review, 15,* 115–45.

Trautwein, U., Koller, O., Schmitz, B., & Baumert, J. (2002). Do Homework Assignments Enhance Achievement? A Multilevel Analysis in 7th-Grade Mathematics. *Contemporary Educational Psychology, 27*(1), 26–50.

Walsh, B., & Dickey, G. (1990). *Building a Champion.* St. Martin's Paperbacks.

Walsh, B., Billick, B., & Peterson, J. A. (1998). *Finding the Winning Edge.* Champaign, IL: Sports Publishing Inc.

Weeres, D. J. (2010). *Just Call Me John: The Leadership Story of John Gagliardi.* North Start Press of St. Cloud.

Made in the USA
Monee, IL
24 March 2026

46726086R00121